Choreography as Embodied Critical Inquiry

Shay Welch

Choreography as Embodied Critical Inquiry

Embodied Cognition and Creative Movement

palgrave
macmillan

Shay Welch
Dept of Philosophy & Religious Studies
Spelman College
Atlanta, GA, USA

ISBN 978-3-030-93494-1 ISBN 978-3-030-93495-8 (eBook)
https://doi.org/10.1007/978-3-030-93495-8

Cover image courtesy of AnnMarie Spexet

This Palgrave Macmillan imprint is published by the registered company Springer Nature Switzerland AG.
The registered company address is: Gewerbestrasse 11, 6330 Cham, Switzerland

Acknowledgments

I would like to acknowledge my invaluable collaborators for the embodied critical inquiry method. First, I would like to thank Constance Echo Palmer and Shannon Stanley for helping me develop and work out the method itself during the beginning stages. I would like to thank Deena Weisberg and Sorsha Masters for helping me iron out the final kinks. And I would like to thank AnnMarie Spexet for all of her emotional and artistic support, as well as for her work doing all of the videography and photography for the book and method development.

Contents

About the Author

Shay Welch is an Associate Professor of Philosophy at Spelman College. She was the 2020–2021 Carnegie Corporation and Rockefeller Foundation Distinguished Research/Creative Scholar. She is Chair of the Association for Feminist Ethics and Social Theory and is a founding board member of Tenure for the Common Good. Her recent book is *The Phenomenology of a Performative Knowledge System: Dancing with Native American Epistemology* (Palgrave Macmillan 2019). Her previous books are *Existential Eroticism: A Feminist Ethics Approach to Women's Oppression Perpetuating Choices* (Lexington Books 2015) and *A Theory of Freedom: Feminism and the Social Contract* (Palgrave Macmillan 2012). She teaches courses on freedom, embodied knowledge, embodied cognition, dance, systemic oppression, ethics, sex, feminism, and Native American Philosophy. Her professional goals are to support and mentor young women of color in Philosophy and to aid the discipline in recruiting and retaining more underrepresented young philosophers.

List of Figures

CHAPTER 1

Dancing: Phenomenology and Embodied Cognition

There is a multitude of movement methods and techniques within dance and the performing arts. With each new groundbreaking choreographer onto the scene, new styles take shape and dancers often choose a specific technique or method to occupy their affections. While not all dance methods or dance theories center on, or are especially interested in, the communicative capacities of dance, this capacity is a powerful dynamic of this particular creative form of movement. While communicating meaning is not always the aim, movement inherently has meaning. Rudolph Laban[1] believed that while personal artistic styles of bodily attitudes are identifiable, and are often deployed, as the means through which to intentionally command the communicative power of dance, it is not the heart or significance of movement. On Laban's account, the significance of movement belongs to the moments of transition between positions that themselves change the expression being presented. According to Laban, "[e]ach phase of movement, every small transference of weight, every single gesture of any part of the body" is germane to the meaning of movement (Laban 1980, 19).

To capture this perception of the communicative force of movement, Laban created and introduced to the dance world a method of notation

[1] Other established, but less universally well-known, dance notion systems include: Zorn, Feuillet, Benesh, Stepanov, and Eshkol-Wachmann. Laban seems to be the one most identifiable, among even mere hobbyists.

S. Welch, *Choreography as Embodied Critical Inquiry*,
https://doi.org/10.1007/978-3-030-93495-8_1

based on movement to help harness and direct the potentialities of meaning in, and through, movement (Davies 2006).[2] Laban, and many other dancer-scholars, find in dance the capacity for a mode of expression for ideas that could not be articulated, but that are also ineffable. For example, Sondra Fraleigh attests that dance is, has, and creates its own meaning and has no word correlates (Fraleigh 1987, 73). The broad mainstream presumption is that dance is affective, rather than cognitive. So while it is a "language", it is not language. While is "speaks to us", it cannot tell us. It "tells a story" but makes no claims. It is said, or one might say, that dance hits us not in the head…but rather in the gut (Reason 2016). Dance, in effect, is immanent. It is supposed as that which fills the interstices of more banal modes of communication and life: personal, raw, emotive, evocative, connecting, affecting…"sensual, sensate, and sentient" (ibid., 89). Many think that we may be able to symbolize the meaning within dance, but we cannot name it (Fraleigh 1987, 73). My philosophical project here is one of embodied knowledge and creative movement; and it sits at this intersection between methodology and (in)effability.

I agree with dancer-scholars that the communicative power of dance is affective, as connoted by the above descriptors; but I disagree that it is also not cognitive and so is not *merely* affective. This is what I mean when I say that dance is acclaimed for its ability to speak to us even though it is not believed to be actually *telling* us something. I hold, along with many others who are working at the intersection of dance and embodied cognition, that dance is both, and equally, affective and cognitive; cognition is affective, and affect is cognitive. Dance can speak to us because, and when, it wants to tell us something. When one claims that movement is both affective and cognitive, they make two sorts of claims. First, they hold that movement is itself a mode of thought and a way of thinking. An embodied cognition account would refute the idea that we must first have thoughts and intentions about moving in our head before we move for that movement to have or denote meaning; this sort of brain-computer model implies that our movement lacks cognitive content on its own. Relatedly, an embodied cognition account of movement would also refute the idea that dance is *merely* the expression of affect, which includes, but is not

[2] Laban's method is grounded in two main components: Choreutics, or the sculptural quality of movement, and Eukinetics, which is the intensity and nature of effort of the activity.

limited to, emotions. On embodied cognition accounts, affect is itself a cognitive process, not an ancillary psychological or biological force that colors or shapes our cognitive states. For this reason, affect is not only part of our cognitive processes, but it is also capable of going beyond the limits of expression. Affect pulled into, and pushed out of, movement can be, even if unintended, strongly communicative. For me, dancing, broadly construed, can tell us what is otherwise ineffable but itself is not "ineffable". In fact, I find movement to be quite explicit, loud, and precise. But I do think that there are more or less effective means of going about how to access and communicate our bodied languaging. Laban and other notation theorists were on to something. They believe that movements can be formalized and described in such a way as to be named. In some sense, I share their motivation. My hope is that I can offer a method through which a mover/dancer may work to intentionally name their body-languaging for intentional purposes.

In this book, I expand on the thinking-in-movement framework that has its roots in the work of Maxine Sheets-Johnstone and extends and develops within contemporary embodied cognition theory. The framework from which I am working holds that the movement expressed and articulated in dance articulates and communicates thoughts and ideas. In my previous work (Welch 2019), I looked to narrative and ritual practices of dance to argue that dance can make claims and tell Truths[3] that are both explicit and objective. But I also believe that dance, and creative movement generally, is capable of asking questions. Thus, my ultimate aim here is to offer a theoretical framework for how critical inquiry can be embodied; and to present one somatic-based dance methodology for creating choreography that demonstrates how this can work. The empirical method I proffer is a movement tool that can be done in its own right for its own sake, engaged in to help one think deeper or differently about a non-movement-based project, or engaged in to help the mover/dancer figure out how to better choreograph a narrative work. More specifically, I explore and analyze the role of embodied cognition and embodied metaphors in mining the body for questions one has about the world and themselves.

My primary purpose is the philosophical purpose. That is, this book is primarily a philosophical venture that is accompanied by a movement methodology as an example for how this philosophical framework

[3] This conception of Truth comes from Native American epistemology.

operates. My philosophical purpose motivates the creation of this method to reveal how to utilize movement to explore our embodied practices of knowing. I argue that it is our creative embodied movements that give us the ability to cognitively and critically engage in critical analysis about the world. I aim to provide an analysis—one that blends phenomenology of dance and a "naturalized" phenomenology of embodied cognition and mind—of what it means to develop, articulate, and express questions that reside in one's cognitive unconscious (or subconscious or nonconscious) to oneself—and to others—through the moving and dancing body. Overall, I hold that through the reading, rather than interpreting, of embodied metaphors, dance can be a clear language available to us. And, more interestingly, it can be a reflective way of accessing the "hows and whys" that we have about the world and our place in it.

I focus on how the phenomenology of dance fits within an enactivist framework. More specifically, I look to various components of enactivist models to clarify how the various parts of an embodied critical inquiry are enactivist. Enactivism is important to a theory of embodied critical inquiry because it can help us better understand how to purposefully engage the body and its knowledges to stimulate meaning-making and meaning-discovery. While dancing is recognized as a mode of artist narrative, there is little research on the specifics of how, and what, this entails theoretically. My primary aim here is to provide some of those details for an embodied critical inquiry. But, some may wonder why an account of an embodied critical inquiry would be of any interest to enactivists. A theory of an embodied critical inquiry is of interest to enactivists because it adds more material content for their empirical research. While cognitive scientists and neuropsychologists have been exploring brain processes through the use of dance, few have done so with a version of dancing that functions in the realm of higher cognitive functions that cannot be explained verbally. My hope would be that enactivists could use this mode of dancing to explore some of the more abstract cognitive functions that still remain somewhat vague within the literature.

To give this project real bite, I conducted a series of movement experiments to produce and reproduce the claims of my argument, which are concomitant with Ann Cooper Albright's intimation that improvisation is a "method of inquiry" (Albright 2019, 25). Through my own work with choreography, and with other choreographers, I have developed an empirical, movement-based practice that can guide individuals through processes of embodied critical inquiry. The methodology relies on the utilization of somatic practices and embodied metaphors through the process of somatic-based movement improvisation. I substantiate my methodology through facilitated improvisational movement exploration to then further develop the embodied notions revealed. This claim about movement is not delimited by, or to, "proper" dance. Generally speaking, embodied critical inquiry is a fact about cognition itself and thus the idea that critical inquiry can be pursued purposefully using embodied practices is rather straightforward. I employ the embodied practice of choreography to set out the framework since it is the most obvious form of intentional movement creation for most people. It is also the embodied practice with which I am most familiar.

I can best explain my work here as one general argument and a series of positive arguments that interweave a theoretical model that substantiates this primary argument. The process that leads me to draw the connections between embodied cognition and phenomenological accounts are heavily descriptive rather than argumentative. I set out a number of other's positions and orient them in relation to one another to show how they operate in tandem to corroborate my argument. My overall argument is that embodied critical inquiry is possible, which can be evidenced from within the embodied cognition schema. Further, I purport that embodied critical inquiry is most successfully pursued and achieved through somatic-based practices of creative movement. My argument rests on the acceptance of the idea that movement and thinking are one. My work is not intended to interrupt traditional cognitivist models, nor does it attempt to defeat any traditional cognitivist foundations. Because this is an embodied knowledge theory, I am strictly in conversation with those who accept embodied knowledge as real and the body as the foundation of knowledge and cognition.

Throughout my work, I engage theorists of various levels of embodied cognition commitments and advocate for stronger versions of embodied cognition. As my methodology shows, dance undoubtedly succeeds as a mode for embodied critical inquiry. What it also shows, is that this success exemplifies and substantiates a strong, rather than weak, version of embodied cognition theory. So, if one were to posit that I am interceding into any philosophical debate, it would be in the debate between different kinds of embodied cognition theorists and not with mainstream philosophers of mind. But I also do not regard my entrance into this conversation as one of interruption or interception. I see my entrance into this debate as one who raises a hand as says "Hey y'all, check this out! Cool, huh!". My interest is more in showing that embodied critical inquiry, as opposed to only embodied narrative, is possible. I bring together various theorists, disciplines, and ideas to make this positive argument. I believe that my claim that an embodied critical inquiry is possible will be mostly uncontroversial among embodied cognition and dance theorists. I believe that actually demonstrating this idea through a dance method is radical and significant. But I hold that the paths of idea-building I take along the way to cement the philosophical analysis of an embodied critical inquiry is rather subtle and highly collaborative.

Because this book is situated within embodied cognition, it will be of most use for those who are somewhat steeped in, or at least peripherally oriented towards, this area. Others who will be at least comfortable with this analysis are philosophers whose work centers on embodied knowledge. This analysis, however, is still of great benefit to dance theorists. While the philosophical work is anchored within the somewhat technical area of embodied cognition, the topics I discuss are topics that also play a pivotal role in contemporary somatic dance theory and dance therapy. Where some of the language may seem inaccessible for those outside of embodied knowledge and cognition conversations, the language of the empirical method should clarify the ideas and provide concrete context to better make sense of the philosophy. Some readers, possibly those in dance theory, may feel as if it would be better for the two prongs of this project to not be demarcated. It may be the case that having parts of the empirical method incorporated into the theoretical chapters as more tangible examples would enhance the accessibility of the theory. Unfortunately, this approach, for me specifically, was not feasible. When I attempted to intermingle the two, the singular sensical nature of the methodology dissolved because each step makes no sense individually. And since no one step makes sense alone, to build discrete steps into the various parts of the theory, which would also require that they be considered out of order, would further obfuscate the ideas within the theory.

I want to be very clear that I do not have any particular school or camp or style of dance in mind when I make these claims. This book is about creative movement and embodied cognition; it is not about dance itself.

When I speak of dance here, I speak of the open-ended creative and artistic and expressive (and inquiring!) mode of movement that could be identified as dancing. Certainly, some dance methods will be more (or less) conducive to the claims I make here in my project. Historical forms of ballet that were mimetic would not fall under my account, but there are plenty of ways in which ballet moves and techniques can be just as communicative as hip-hop, contemporary, or modern. And all kinds of dance that are called so only out of sheer social generosity qualify too. What I ask the reader to do is to think broadly about the powers and abilities and capacities of dance as an expressive, artistic, querying, creative activity. Certainly, some dance has no intention of communicating or querying about anything at all, and that's alright. The method I put forward is flexible and necessarily incomplete. It may work for some dance forms more than others. My method is simply intended to help guide movers through embodied practices of critical inquiry, which I conceive of as a purposeful and directed mode of generating questions about oneself and the world. I *do* argue that we can engage in processes of critical inquiry through movement and choreographing practices at the embodied cognitive level. I do *not* suggest that all dance is of one kind and involves, or should involve, critical inquiry or narrative force. My method is not rigid and particular; it is open and fluid and concerned with, and invested in, the potentialities of creative cognitive movement that have yet to be fleshed out in philosophy.

I should very strongly emphasize that my experience with "dance" is actually in aerial, since I am an aerialist and not a dancer. I have no professional dance training and work strictly with artists in the circus and experimental arts community. I am writing this from a first-person account of my experience with aerial—otherwise known as "sky dancing", which also includes wall dancing—so my understanding of dance is broader and might, in some places, have more in common with commercial and circus performing artists than with academic and fine arts dancers. But while I am not classically trained, the artists I work with on the method have classical dance training. This is why I chose them to help me with this project. They were not my "subjects"; they were crucial collaborators in the process development, and they were able to help me hobble along where my lack of formal education revealed itself. I hope to bring a fuller appreciation of multifarious movement arts to the discipline of philosophy, and I hope that I can get my readers to join in on this jubilant shenanigan.

As a final introductory remark, I want to emphasize that I regard this project as form of public philosophy—not the theory per se, but rather the actual movement methodology for embodied critical inquiry. I consider this public philosophy because I want my method to be of use to public artists and communities in public spaces. I want this project, not particularly in these specific words, but in its purpose, to have applicability to teachers and students in local studios who want to learn how to be more creative and purposeful with the construction of their narrative pieces. I have begun to share these ideas with choreographers and local performers through lectures, workshops, and consulting as a means of contributing to community performance arts' ability to engage in social commentary. I formalized this experimental methodology of cognitively informed movement exploration and development as embodied critical inquiry to help other choreographers and movement artists deepen their abilities to use improvisation theoretically in their production development processes. While this method can be used for nearly any narrative theme exploration, my primary motivation for developing this method is for it to be used to choreograph about conceptual and social justice issues, such as racism/police brutality, colonization, gender, sexuality, etc. One of my motivations to develop this theory and method comes from the social justice work in circus. My goal for this project, particularly in the circus community—as a mode of public philosophy—is to help give these performing artists movement-based practices to help them think through some of the larger social justice themes they want to choreograph about so that they can connect to their own personal perspectives on, and interests in, social justice.

Phenomenological Meaning in Dance

Contemporary dance theory has advanced our understandings of the depth of meaning in dance. Questions are not simply limited to "what is dance?", "is dance art?", or "what does dance mean?". Dance theorists and dance philosophers have made great strides in evidencing the ability of dance to create and communicate meaning. Now, when we think about dance, we can think about the capacity dance has as a mode of meaning-making. Dance is now conceived as a proper form of expression and an effectual and valuable vehicle for narrative. Dance philosophers, specifically, have revealed the depths of the power of dance. The dance philosophers whose significance to the field, as it relates to the phenomenological foundations of my project, merit focused attention are Maxine Sheets-Johnstone,[4] Sondra Fraleigh,[5] and Susanne Langer.[6]

For Langer, too, dance belongs with the ineffable, the affective, but not the cognitive. She explains that:

> Whatever there is in expression that will not take the impression—directly or indirectly—of *discursive* form, is not discursively communicable or, in the strictest sense, logically thinkable. It is unspeakable, ineffable; according to all serious philosophical theories today, it is unknowable...Yet there is a great deal of experience that is knowable, not only as immediate, *formless, meaningless* impact, but as one aspect of the intricate web of life, yet *defies discursive formulation*, and therefore verbal expression: that is what we sometimes call the subjective aspect of experience, the direct feeling of it. (Langer 1957, 22, italics mine)

For Langer, art is expressive, yet it is not self-expression. For this reason, dance is an expressive art that yields a virtual symbol. Dance presents apparitions of the subjective feelings expressed that dissipate the moment they are performed. The symbols of dance are virtual in that they are real qua perception, but they exist nowhere. What exists as the tangible facets of dance are the movers and the forces they generate to enliven the apparent dynamic images that convey the subjective feelings. The dynamic images are objective symbols, forms rather than feelings, because they are representations of the subjective feelings being portrayed. That is, the subjective feelings rendered in and through the dance are not the actual subjective feelings of the dancer in that moment. Rather, the dancer is

[4] See: Maxine Sheets-Johnstone. *Illuminating Dance: Philosophical Explorations*. Sheets-Johnstone, Maxine (ed.) London: Associated University Presses, 1984; *The Roots of Thinking*. Philadelphia; Temple University Press, 1990; *The Corporeal Turn: An Interdisciplinary Reader*. Charlottesville: Imprint Academic, 2009; *The Primacy of Movement (Expanded)*. Philadelphia: John Benjamins Publishing Co., 2011.

[5] See: Sandra Fraleigh. *Dance and the Lived Body: A Descriptive Aesthetics*. Pittsburg: University of Pittsburg Press, 1987; In *Moving Consciously: Somatic Transformations through Dance, Yoga, and Touch*. Fraleigh, Sondra (ed). Urbana: University of Illinois Press, 2015; *Moving Consciously: Somatic Back to the Dance Itself: Phenomenologies of the Body in Performance*. Fraleigh, Sondra. Urbana: University of Illinois Press, 2018a: 11–26.

[6] See: Susanne Langer. *Philosophy in a New Key: A Study in the Symbolism of Reason, Rite, and Art*. Cambridge: Harvard University Press, 1947; *Feeling and Form*. New York: Charles Scribner's Sons, 1953; *Problems of Art: The Philosophical Lectures*. New York: Charles Scribner's Sons, 1957; *Mind: An Essay on Human Feeling*. John Hopkins University Press, 1967.

representing subjective feelings felt earlier through her movement via the construction of a dynamic image with the actual present dynamic forces. This is how the expressions of dance can be *virtual* symbols; they are an objective presentation of a subjective reality (9, 21). She believes that the reason why people get confused about what dance is, and what dance does, is because it works on us through our intuitive capacities that allow us to understand one thing in terms of another (20); that just is what symbolism is and does—it is and does metaphor.

Fraleigh is in accord with Langer on the claim that the essence or meaning of dance is not identical with the motions. Yet she takes the meaning of dance deeper; she moves from dance as an expression of feeling to an understanding of dance as consciousness. What this means is that she takes a step back from the performance of the dance by the dancer to its process of creation by the choreographer. One can think of Fraleigh as taking us to the intuitive capacities themselves that Langer tells us are what allows us to make sense of the symbol. For one could say that it is our intuition that moves us to recognize that with which the idea can be symbolized. According to Fraleigh, our intuition consists of sense and meaning (Fraleigh 2000, 60). While intuition is prevalent in her work, her main focus is on the phenomenology of the embodied knowledge we have pre-reflectively and that can be accessed through our soma-sensory awareness; which she identifies as our core consciousness (Fraleigh 2000, 57). The expression of these feelings is our consciousness extended, or as she puts it, "expression is consciousness lived twice" (59). She states that:

> [The] poetic and explanatory demonstrations of experiential affective phenomena bring to our attention the connections they have assimilated by representing in an intelligible form the original intuition. (60)

For this reason, dance does not only *express* some felt meaning; we, ourselves, attribute meaning to it, and meaning inheres in it, beyond the meaning it imparts when we witness it.

Maxine Sheets-Johnstone incorporates Langer's use of the virtual and follows Fraleigh's grounding of the meaning of dance in phenomenology. Most importantly, she developed the thinking-in-movement framework. While she draws heavily on Langer who argues that dance is not philosophy, I purport that Sheets-Johnstone's TIM model is the pinnacle model of actively performing embodied philosophy. Thinking-in-movement is exactly what it sounds like—it is a mode of cognition in which there is no dualism between the thinking and the doing. We do not first think and then move, we do not move and then think. She explains that "the body inhabits movement in the literal sense of living it" (Sheets-Johnstone 1981, 402). She expounds:

> The movement which I actually create at any moment is not a *thing* which I do, an action which I take, but a passing moment within a dynamic process, a process which I cannot divide into beginning and endings…Thinking in movement is a perpetual dissolution and dilation, even a mutability, of here-now movements and a moving present. In fact, I am creating the dance moment by moment in the possibilities of movement I discover in moving and in the possibilities of moving I discover in movement. (405)

The movement itself is the thinking itself. For those who embrace a cognitivist philosophy of mind framework, this sounds…cute. But for many of those who "move", there are few things more obviously true than this idea. Thinking-in-movement is grounded in what she terms a bodily logos, which is a sensorimotor intelligence that has an underlying rationality and logic (403); it is neither pre-rational nor pre-reflective since the thinking is in the moving.

Thinking-in-movement is not just *a* philosophy, though one can see how it can be one way of moving and living in the world existentially. Thinking-in-movement *is* philosophy. The utilization of movement improvisation, whether in the studio or on a crowded street, is a fleshed mode of manifesting and engaging phenomenological possibilities—both embodied and conceptual—in the world and following them to the creative, but logical, ends. Our bodies are, and make, meaning as they swivel and move, cutting lines across spaces that connects one claim of existence to another. Movement, of arm or breath, is fully open-ended in a way that cannot be predicted or pre-predicated. It is fluid and in flux, ripe for invested exploration of the truths within, and upon, the skin of ourselves and the skin of the world. She would call this a kinetic happening (ibid.).

One reason why I think that Sheets-Johnstone is able to pull on Langer and proffer a TIM framework is because Langer seems to be referencing performed choreography and Sheets-Johnstone—at least in her work on TIM—seems to be referencing improvisational movement. As she rightly points out, the practice or engagement is not the creation of a dance—it is the process of the dance. Improv practices, loosely defined and broadly conceived, are neither plucked from a pile of movement vocabulary nor polished panache from a prefabricated narrative. Generally speaking, improv is a creative process of TIM where the transitions between movements are, themselves, the preinventive structures of thought; this idea seems to align with the moments of meaning that Laban recognized. If we were to think of this in a Bergsonian way, we could say that it is a movement praxis that creates its own future as it absorbs its own past. It consumes and enfolds itself as it moves ahead along the logical path that it *must* go. A second reason is that Sheets-Johnstone is positing the mover/thinker/perceiver as one in the same person. Langer talks about dance and art as it is made by one to display before another. This relationality of art inserts a chiasm between maker and perceiver that can disappear when the two become one.

Phenomenology and the Embodied Mind

When it comes to dance philosophy, dance phenomenology sits closest to embodied theories of cognition.[7] Broadly speaking, phenomenology is the philosophical framework that asks us to step away from consciousness to see how consciousness actually works. In a sense, then, it is pragmatic in an existential sense; and it is considered more of a point of view rather than a theory. In other words, we generally take the world for granted and phenomenology asks us to investigate our existence as already being thrown into a world with others. To understand the how and the why of our lived reality, what may or may not be "out there" matters not. What does matter, though, is our experience. The phenomenological perspective is one of radical reflection, which is a "creative operation" we undertake to take on our unreflective experience (Morris 2008, 112). Certainly, Husserl's notion of the lifeworld, which is the world we live in and not the world in front of us, plays a significant role, especially within Fraleigh (2018). But broadly speaking, Merleau-Ponty functions as the cornerstone to most of the phenomenology that grounds embodiment frameworks because he prioritizes the body as the site, and creator, of meaning and experience.[8]

[7] See also: Philipa Rothfield. "Differentiating Phenomenology and Dance". *Topoi* 24 (2005): 43–53.

[8] See: Hass, Lawrence. *Merleau-Ponty's Philosophy*. Bloomington: Indiana University Press, 2008. Merleau-Ponty, Maurice. *The Visible and the Invisible*. Lefort, Claude (ed.). Evanston: Northwestern University Press, 1968; *The Phenomenology of Perception*. Landes, Donald (trans.). New York: Routledge Press, 2012.

Because dance is a phenomenological-cognitive activity-experience, it is helpful to understand how embodied cognition theorists have drawn the connection between movement and phenomenology.[9] At its core, a phenomenological account of consciousness starts from the first-person subjective experience to provide descriptions of experience and how experience is perceived. Intentionality, under a phenomenological frame, is concerned with the intentionality of perception and how our interpretations of experience are dependent on our goal or context. Under phenomenology, perception is already recognized as meaningful and enriched by the possibilities of our embodied experiences in the lived world; it is also informed by our bodily abilities and skills (Gallagher and Zahavi 2012, 8). Such an account of perception is what grounds the phenomenological framework as inherently dynamic since its core components are themselves dynamic. Nothing about our understanding of the world comes from static pictures of consciousness and experience. Perception is intentional and this intentionality is dependent upon the dynamic nature of our bodies in relation to the world and others. Therefore, phenomenology does not only yield a dynamic depiction of the mind, but it also yields an account of dynamic cognition that is a product of dynamic embodied movement and direction. In short, cognition is action and thus movement is constitutive of cognition. This dynamic system lays the descriptive and practical foundations for a realistic methodology for embodied critical inquiry.

Because consciousness is front and center in phenomenology, the study and practice of somatic bodywork practices, broadly construed, unlocks a distinctly embodied phenomenological reduction to help us bracket presuppositions about our experience in the world. Fraleigh draws on somatic movement as a means of accessing and extending consciousness (2015b). This is because, as Husserl posits, our intuition gives us access to the structure of consciousness; intuition grants a pure seeing of consciousness rather than inferring something about it (Gallagher 2010, 24). And as I show later, intuition, whether as affect-cognition or method, is fully embodied and ties directly to our soma. Soma is a kinesthetic affectation allied with the maintenance and regulation of our "minded body" (2015). Soma is the seat of our embodied conscious awareness in that it grounds

[9] See also: Gleen Caruthers. *The Feeling of Embodiment: A Case Study in Explaining Consciousness*. New York: Palgrave Macmillan, 2019; Ted Toadvine. "Phenomenology and 'Hyper-Reflection'". In *Merleau-Ponty: Key Concepts*. Diprose, Rosalyn and Jack Reynolds (eds.). Stockfield: Acumen Pub, 2008: 17–29.

the perception and self-perception that informs our consciousness (xxi). That we have a minded body, as she puts it, means that we embody the values of our experiences before our experience can even arise to the level of explicit awareness (2015, 30).

For Merleau-Ponty, the standardly accepted dualism between body and mind is bizarre. For him, there is no discernable distinction between the "I" and the body since it is through the body that we "I" via our "I can". But even more poignant for the understanding of an embodied critical inquiry, our body actually "outruns" the awareness of our subjective experience (Weiss 2008, 133; Busch 2008, 34). Our body schema, and the movement it ignites, generates and distributes meaning; movement engenders meaning because it is not reducible to the body (Morris 2008, 118). But also, our movement also unites us—our embodied consciousness—with the outside, such that our inside is part of the outside and the outside is part of our inside. Merleau-Ponty refers to this peculiar relationality as "flesh". Flesh is that which binds us all to each other and to all things. Flesh is the glue that holds us, and it, all together—"the narcissistic intertwinement of subject and object" (Evans 2008, 188), as each one is the side of the other. Flesh is also important to Sheets-Johnstone (1984, 2009) given its relationship to skin; the animate surface sensitivities of the skin are what connects us with, and weaves us into, the flesh. She eloquently remarks:

> Skin holds us together; flesh binds us together. Skin holds us together and grounds our knowledge of ourselves and of the world; flesh binds us together and grounds the possibility of an enlightened and reciprocating humanity that, celebrating the diversity of life, celebrates also its commonalities, and that taking from the flesh of the earth, takes care also to protect and preserve it. (147)

The role of the body and the notion of flesh in phenomenology are requisite for an intelligible understanding of an embodied inquiry. Flesh is what binds our embodied consciousness to others and the world and through which our skin sops up the vast array of knowledges that become embedded in our bones. But movement moves meaning because it moves us across, and into, the flesh (Morris 2008, 118). Or as Merleau-Ponty says: "…because we are the world, we are *condemned to moving*" (PP: xix; Busch 2008, 32). Thus, our body is not a passive receptacle of meaning and knowledge. Our body moves us into meaning and actively takes up the knowledge into an embodied residence. Thus, it is through our skinned movement through flesh that we absorb knowledge, and back through our skin through our movement that we can expel it.

The significance of the body, embodied subjectivity, and this notion of the flesh, which pulls together one, and others and environment, in phenomenology has not gone unnoticed in other subfields and disciplines. Phenomenology intersects Pragmatism, Feminist Philosophy, Critical Race Theory, and Indigenous Philosophy. These intersections are not surprising and are rather expected. But the domains in which the priority of the body is surprising is in the Philosophy of Mind and Cognitive Science. Following the work of Dewey (1931, 1958) and James (2000), Bergson (2007) and Whitehead (1978), and Eugene Gendlin (1992), there is a sizable shifting in the consideration of the body itself in perception and cognition—beyond the role of a passive receptacle for sense data and qualia, to that of an embodied realism.

Alva Noë explains that the world makes itself available to us through our interaction with it; and what we perceive becomes a function of what we know how to do.[10] Thus, all perception is intrinsically active and thoughtful (Noë 2006, 3). Similar to the phenomenologist, Noë posits that the world makes itself available to us; we not only purposefully select what to direct our attention and focus towards, we actually have to "reach out and grab it" with our sensing organs. Our capacities to look and move and smell make it possible for us to perceive; but effective perceiving cannot happen unless we develop these capacities into sensorimotor knowledge (75). Noë purports: "In actively encountering the way in which how things look varies with movement, we *directly* encounter how things are" (85). The more we are able to go through the processes of developing our bodily sensorimotor skills in relation to our other perceiving organs and our environment then we are able to directly perceive more of the world, particularly if we understand perception as an action that is meaningful and that makes sense. Our ability to see things for what they are is entirely

[10] In my previous work, I have labeled Noë as an Enactivist. This is not wholly true even though much of his theory is about demonstrating how perception-as-action leads us to enact the world. There is considerable overlap between embodied theorist, such as the functionalists, the extended mind theorists, the linguistic embodied theorists, the low-key enactivists, and the radical enactivists. These distinctions were either unrecognizable to me in my earlier work or were not clearly demarcated in the material. In this work, I try very diligently to keep these distinctions correct and separate but they are all very intimately intertwined so either there will be almost no difference or I will still be unable to fully track them.

dependent upon our ability to move around, and with, them. As an example, he explains that we know that—and we perceive—plates as round even when they appear to be elliptical in shape because we have a past of interacting with plates and we come to know that round items will appear elliptical if we move in certain angles or distances. We see that things are what they are accurately because we are seeing them "from here", wherever here may be in relation to the object, or objects in relation to each other and oneself.

Noë's phenomenological account of perception as action qua our bodily sensorimotor skill grounds many embodied mind theories that prioritize the body in cognition.[11] According to embodied cognitive theory, meaning is phenomenological and stems from embodiment in that it comes together for us through nonconscious and mostly unaware bodily perceptions of space, movement, and environmental qualities that constitute our experiences (Johnson 1987, 2007). Movement, specifically, grounds our ongoing connection to, and interaction with, others and the world; it is what keeps us in touch with the world (Johnson 1987, 20). However, he tells us, because meaning is born from unconscious embodied perceptions and movements, its role in the process of worldmaking becomes invisible.

Johnson intimates that the world as it presents itself to us, and our situatedness in that contextualized world, is minimally meaningful and requires our active embodied participation in, and with, it to extract and construct the kind of meaning that makes life intelligible and meaningful. Our contextualized positions are a field of possibilities and opportunities; and as we think and act, we create and structure meaning by creating connections. This then leads to further opportunities to explore and inquire (2007, 265). Initially, meaning arises from embodied movement and interactions that are later extended metaphorically in our linguistic and conceptual mappings. A body image schema develops when our sensorimotor experiences track repeated patters and relations. The resultant image schemas are what give our broader experiences shape and meaning, as well as serving as models and modes of reasoning, insofar as the repetitions generate neural mappings that eventually constitute what gives rise to abstract thought. Through our embodiment, we come to have understandings of, and meanings for, experiences of ourselves and others.

[11] See also: Summa, Michela. "Enacting Perception: The Relevance of Phenomenology". *Paradigmi: Rivista di Critica Filosofica* no. 2 (2014): 97–115.

Ultimately, without our body's capacities to act—to move, perceive, manipulate, and engage—we would have no source from which to imaginatively draw ideas, induce, or infer.

Enactivism positions action at its center and describes cognition as a function of our goals. Johnson rightly and explicitly centers movement in embodied cognition. He argues that "[m]ovement thus gives us knowledge of our world and, at the same time, reveals important insights about our own nature, capacities, and limitations" (Johnson 2007, 19). Movement qua sensorimotor skills is a necessary feature of cognitive processes, such as planning, predicting, reasoning, and generating behavior. Once one learns a particular skill via movement that then manifests as a sensorimotor contingency, that sensorimotor contingency could, at a later point, predict whether some action will be rewarding for our designated goal. That is, our sensorimotor contingencies are, themselves, capable of providing the basis for action plans to achieve some specified goal (Engel et al. 2015, 7). For example, movement impacts vision since visual awareness draws on previously established sensorimotor contingencies to properly "see the scene" in a way needed to successfully achieve some goal (9). Action and movement drive our phenomenological states since mind is conceived of as an achievement rather than as a given.[12] Ultimately, Johnson argues, life, meaning, and movement are connected (Johnson 2007, 19).

Johnson often gives what I find to be the most illuminated account of movement, specifically, in embodied cognition and cognition as action—especially because he pulls from Sheets-Johnstone so frequently. He cites her as highlighting the importance of kinetic spontaneity for living in, and making sense of, the world (ibid.). Meaning is a function of our embodied actions in the world and these interactions are either initiated via actual movement or the sensorimotor contingencies constructed from prior movements. Whether currently in motion or not, meaning initiates through movement. He rightly explains that moving keeps us in touch with our environments, relations, and contexts. He also draws on Langer to shore up his account of the embodied meaning view in the context of music.

[12] See also: Victor Loughlin. "Sensorimotor Knowledge and the Radical Alternatives". In *Contemporary Sensorimotor Theory: Studies in Applied Philosophy, Epistemology, and Rational Ethics*. Bishop, John and Andrew Martin (eds.). Springer International Pub., 2016: 105–116.

> To hear music is just to be moved and to feel in the precise way that is defined by the patterns of the musical motion. Those feelings are meaningful in the same way that any pattern of emotional flow is meaningful to us at a pre-reflective level of awareness…Langer is saying that when we are actively listening to music, we imaginatively enter into its "motion," experiencing all of the ways it moves, swells, hops…This *musical* soaring, floating, or falling is experienced by us as *our felt flow* of experience. We feel it in our vital, tactile-kinesthetic bodies. (239)

It is no wonder, then, that scholars should look to dance as the paradigm mode of purposeful exploration of this embodied givenness for the intentional aim of mining, creating, and communicating meaning.

Embodied Metaphors and Enactivism

Because phenomenology and embodied cognition revolve around the lived body for understanding cognition and consciousness, the role of metaphor, which has always been pivotal to our conceptual linguistic experiences, takes on a new role—or rather, puts on a new guise…a body. Berndt Sellheim draws on the conception of the flesh to establish the link between Merleau-Ponty and embodied metaphor. The metaphor itself is from the physical flesh to the conceptual flesh. He claims that the metaphorical nature of flesh forms abstract cognition and hyper-reflection that cannot be reduced to words; this follows the fact that the subject cannot be fully reduced to its physiological processes—the subject is non-reductive. It is the inherent link between the body and language that drives Merleau-Ponty's conceptual metaphor of flesh (Sellheim 2010, 267). Unfortunately, we presume that outside of phenomenology, much of our strict theoretical and scientific posturing is free from poetic, imaginative metaphor because embodied metaphors are concealed. Or rather, they are *so* visible as to be rendered invisible. The recognition of the relationship between language and the body in phenomenology helps us put to bed the artifice of a fully disembodied mind.

From here, there is a swift move from metaphor about the body in phenomenology to phenomenologically grounded embodied metaphors in cognitive linguistics. George Lakoff and Mark Johnson developed much of the core framework of embodied metaphors that anchors many models of embodied cognition (Lakoff and Johnson 1980; Lakoff 1990; Johnson

1987, 1993, 2007, 2017).[13] In their collaborative and individual works, they illuminate why it is impossible to think about our subjective experiences without these embodied metaphors; and any attempt to do so would result in impoverished understandings of the world and our existence in it. Conceptual embodied metaphors originate in our embodied image schemas when we pair up a physical experience with a subjective experience. These embodied metaphors constitute our primary embodied metaphors that carry our embodied logics and are then extrapolated and applied to the world conceptually. The repetition of patterns between physical sensation or orientation and the conceptualization of the experience produces neural maps from the physical to the conceptual. In the "affection is warmth" metaphor, warmth is the source domain and affection is the target domain. When one experiences affection, the neural cluster for warmth becomes activated. Our use of conceptual embodied metaphors occurs naturally and automatically and become ingrained incognito into our linguistic understanding of the world as they become systematized through social and cultural use. They are made invisible insofar as most of them are used systematically, and the metaphors themselves are experienced systematically.

Lakoff and Johnson are often critiqued by enactivists for placing too much emphasis on neural maps and mirror neurons (Gallagher 2017, 33).[14] Enactivism is committed to the stronger philosophical and phenomenological conclusions of embodiment—even more so than Noë.[15] At

[13] See also: Steven Winter. *A Clearing in the Forest: Law, Life, and Mind*. University of Chicago Press, 2001. Zoltán Kövecses. *Metaphor in Culture: Universality and Variation*. Cambridge: Cambridge University Press, 2005.

[14] In my previous work, I rely almost exclusively on mirror neurons in the embodied metaphor model to substantiate my arguments. I now see that there is a tension but I do not see it as unresolvable. I will attempt to relieve some of the tension here but am aware that I will not be able to do so completely. Yet it should be noted that there are multiple positions on the spectrum from weak embodied cognition to radical embodied cognition.

[15] Noë would be considered a weak embodied cognition theorist who holds that the fruits of our sensorimotor, action-driven perception get moved offline in the form of mental representations for later use. The most radical claim of the most radical version of enactivism is that there are absolutely no mental representations, zero, zilch, and that cognition and consciousness is naturalized. In other words, the brain, cognition, and consciousness are not solely in the brain and are more prominently functions of our sensorimotor coupling with the world. See: Daniel Hutto. "Knowing *What*? Radical Versus Conservative Enactivism". *Phenomenology and the Cognitive Sciences* 4 (2005): 389–405. Richard Menary. "Introduction: What is Radical Enactivism". In *Radical Enactivism: Focus on the Philosophy of Daniel H. Hutto*. Menary, Richard (ed.). Philadelphia: John Benjamins Publishing Company, 2006:

its base, enactivism holds that perception exists specifically for action, and our action constitutes most of our cognitive processes, where cognition is both normative and affective. Shaun Gallagher refers to this framework as a kind of nonreductive naturalized phenomenology. Enactivists purport that there is no need for mental representation since action and perception can handle everything (14). A weaker way to express this notion is to say that enactivism maintains that mental representational structures and mechanisms are not the best tools to explain the dynamic world-making cognitive processes that constitute lived experience and consciousness (Gallagher and Rucińska 2021, 2). Enactivism is considered a nonreductive form of naturalism because enactivism is a non-representational account, but it also does not boil consciousness down to biological processes. Cognition is a function of the dynamic relationship between our brain, body, and environment. The dynamic system of enactivism is the body-brain-environment sensorimotor coupling itself. Several theorists, including Gallagher and Hubert Dreyfus (2014), explicate how Merleau-Ponty's phenomenology is an enactive non-representationalist schema.

It is important to note that while strong and radical enactivists regard Lakoff and Johnson as extending a weak embodied cognitive theory, they themselves reject this designation. Their rejection of this label is significant because my arguments rest on an account of embodied metaphors that is also enactivist. I attempt to relieve some of the tension here but am aware that I will not be able to do so completely since there is no definitive scientific research that eradicates mental representations. Yet it should be noted that there are multiple positions on the spectrum from weak embodied cognition to radical embodied cognition. There are enactivist positions that permit of weakly neuronal action-oriented representations (103), especially for higher level modes of cognitions. "*If there were any such thing as an enactive representation, clearly this would be it*" (87, italics mine; Rucińska and Aggerholm 2019).

Enactivism relies on affordances as explanans for action-centered cognition. Affordances follow from our sensorimotor schemas, or sensorimotor contingencies, and allow us to enact the world through our sensorimotor coupling with it. Affordances replace representations in an enactivist view; they are dynamic and relational in a way that even action-oriented representations cannot be (15). Given our specific sensorimotor body schemas,

1–12. Some enactivists reject the most radical version. Noë also lacks an affective component to his account of perception, which is deleterious.

the world affords us varied possibilities for action in every context. There are things that the world can offer humans that it cannot offer other-than-human persons, and vice versa, consequent of our body shapes and mechanics. We explore our affordances via imagination in specific contexts to further enact the world for us and create meaning. Imagination, as a higher-order cognition, is itself an action that occurs within action, not before. We develop and use our imagination faculty through pretense and pretending—otherwise known as playing make-believe. Most of this is achieved through enacting embodied metaphors (193; Gallagher and Lindgren 2015). The fact that we enact the world through affordances available to use through our sensorimotor body schemas, and ability to engage imagination through enacting metaphors, is another reason why I do not feel hazardous situating embodied metaphors concomitant with enactivism for my argument.

Dance and Embodied Metaphor

Traditional metaphor theory has been no stranger to dance and dance has been no stranger to traditional metaphor theory. And well before embodied metaphor theory, scholars such as Nietzsche, were combing both dance and metaphor to describe the process of doing philosophy. One such theorist, aside from Nietzsche, who is quoted repeatedly when thinking about the connection between philosophy, metaphor, and dance is Paul Valéry. Most notoriously, he avers that:

> What is a metaphor if not a kind of pirouette performed by an idea, enabling us to assemble its diverse names or images? And what are all the figures we employ, all those instruments, such as rhyme, inversion, antithesis, if not an exercise of all the possibilities of language, which removes us from the practical world and shapes, for us too, a private universe, a privileged abode of the intellectual dance? (Valéry 1983, 65)

And in order to explicate why motion, rather pose is the loci of meaning in dance, Trevor Whittock tells us that:

> This is why [conventional] metaphors are so useful to artists: they rejuvenate our perceptions. They make us see things afresh in their uniqueness, whereas too often we stop looking closely at things when once we have categorized them. (1992, 245)

While Whittock's subject here is dance itself, it is clear to see how this reflection bears a strong resemblance to the schools of thought that demarcates philosophy as *process*.

Langer also has an account of metaphor, which is a relatively established approach. For her, metaphors are what our imagination grabs hold of when we have ideas and experiences that are ineffable. When we fail to find the appropriate language to articulate ourselves, we plug metaphors into our linguistic gaps. Therefore, Langer holds that metaphors themselves are not discursive since they are not language, but merely are ideas expressed through language (Langer 1957, 23). Since art also does not state ideas but rather expresses them, metaphor and art appear to be intrinsically linked. Both art and metaphor give content to the gaps in our propositional language concerning experience and can do so because our knowledge is wider than our discourse (ibid.). Moreover, neither art nor metaphor are abstractions—both are apparitions, virtual.

Langer claims that a good work of art makes it seem as if it truly contains that which it represents: "The congruence is so striking that symbol and meaning appear as one reality" (26). This is also true with respect to embodied metaphors. However, embodied metaphor theory diverges from Langer's understanding of metaphor since she regards them as plug ins, whereas embodied metaphors have linguistic-cognitive content. Where metaphors are not discursive for Langer, the same would not be true of embodied metaphors since embodied metaphors are embodied language. When Langer claims that symbolic logical forms are not visible in art, such as dance, she misses the body and kinetic logics, set out by both Johnson and Sheets-Johnstone, that does make some metaphors—embodied metaphors—discursive and visible. Because the metaphors the body wears are not abstracted or extended representations, this makes it consistent with Langer's claim that art, such as dance, is non-representational (125). But this again draws out another chasm between the two since Langer further argues that if something is non-representational then it can only be an expression of feeling and lacks idea content (ibid.). Conversely, Lakoff and Johnson point out that our metaphors are not free of content since they are the target domain of the actual source domain. They argue:

> Imagine a concept of love without physical force—that is, without attraction, electricity, magnetism—and without union, madness, illness, magic, nurturance, journeys, closeness, heat, or giving of oneself. Take away all of

> the [embodied] metaphorical ways of conceptualizing love, and there's not a whole lot left. What's left is the literal skeleton.... (Lakoff and Johnson 1980, 72

But the meeting between Langer and embodied metaphors is not entirely lost. As an aside, I focus here on Langer, since Sheets-Johnstone, among other dance scholars, draw on her work for their own. The avenue through which one can access the direct communicative (not expressive) power of embodied metaphor in dance is through Langer's positing of gesture as having linguistic content.[16] Langer argues that gesture is discursive because gestures are the abstract symbols of the feelings and intentions that get expressed through them (Langer 1983, 29). They are also actions insofar as they are a kinetic experience. (Langer 1953, 173). Gesture upgrades (or downgrades?) from daily parlance to art when the dancer imagines the expressive gesture to perform it (174). Gesture loses its symbolic form when it is being used in dance because it gestures towards what was imagined to be expressed in performance. In performance, gestures are illusions and transmute in the virtual.

> The bodily movement, of course, is real enough; but *what makes it emotive gesture*, i.e., its spontaneous origin in what Laban calls a "feeling-thought-motion", is illusory, so that movement is "gesture" only within the dance. It is *actual movement*, but *virtual self-expression*. (1983, 31)

But embodied metaphors in dance are more than colloquial gestures that work alongside verbal communication. Embodied metaphors are versions of linguistic gestures and, thusly, are not virtual just because they are constructed through the creation process for the later purpose of performance. When embodied metaphors are performed, they lose no content between the time that they emerged through, from, and with feeling, and when they are communicated through, and with, bodily movement in performance.

In addition to myself, the leading scholar on embodied metaphor and dance is performance philosopher Einav Katan-Schmid (2016; 2017). Quite in line with the account of embodied cognition put forth by folks like Lakoff and Johnson, Katan-Schmid explains that the cognitive experience

[16] This claim is not unique to Langer. It is a widely accepted claim within philosophy of language.

of meaning and meaning-making of dance resides directly in its physicality. Her analysis demonstrates the depth of embodied cognitive metaphor theory by demonstrating how dance is, in essence, the manifestation of meaning qua movement. That is, dance does not represent the meaning of something, it articulates meaning, specifically through the use of somatic precision in, and through, movement. Because dance does not represent meaning, but rather holds, shapes, and lives meaning, Katan-Schmid posits that dance must be danced for the ideas to be manifested and realized (2016, 16). She identifies dance as a communicative expressivist mode constituted by and through the perceptual process of generating actions with meaning. It is through such perceptual processes that dance both expands the cognitive capacities of the dancer and makes dancing cognitively intelligible to viewers. According to Katan-Schmid, the creative perceptual process of engendering meaning through movement exists at the point where sensuality meets intentionality. Fundamentally, active perception in dancing necessitates that dancers gaze into the sensual sources in, and of, their bodies and attempt to interpret understandings of those sources in relation to movement (50). She explicates that since qualities of movement originate in embodied felt experience, dancers must intentionally pull from sensation and direct them into their created dancing gestures (191). The physical actions that are utilized, both cognitive and gestural in nature, actualize the sensations and make them comprehensible—through enactive embodied metaphors.

While the understanding of embodied metaphor, especially in dance, was not available until very recently, the lived experiences of embodied metaphors are what underlie and substantiate Sheets-Johnstone's analysis of TIM (1981; Merritt 2015, Welch 2019). She argues that thinking-in-movement relies upon our kinetic bodily logos. A kinetic bodily logos manages and shapes the embodied creative patterns of thought—it structures "motional thought", as she puts it. We can see the origins of the embodied cognitive theories in her arguments that animation is the primary and original natural being of a body. She holds that movement provides a ready perceptual kinetic susceptibility for life (1999, 494); she argues that "[a] kinetic intelligence is forging its way in the world, shaping and being shaped by the developing dynamic patterns in which it is living" (489). Sheets-Johnstone pulls on improvisational dance to explain how embodied practices of thinking qua kinetic bodily logos engenders kinetic meaning. She espouses that dance is a means through which to explore the world in movement to uncover the extent to which movement creates, and provides, opportunities to create meaning anew. Rather than wonder about the world through words, she says, dancing allows one to wonder about the world directly through movement since thought itself is kinetic (486). The kinetic bodily logos of thinking-in-movement is another way of conceiving of the preverbal or nonverbal nature of movement as procedural meaning-making and communicative action. In fact, she rightly claims that while verbal prose may frequently be ambiguous, embodied dynamics are precise. This is because verbal language is not, and does not, constitute experience. Therefore, attempts to verbalize experiences obscures the fine qualitative and affective constituents of experiences that make them rich and unique to the individual. This is also why we find so many of our abstract concepts filtered through embodied metaphors. To give substance to our language in a way that can capture experience requires us to revert back to thinking about experience with, and through, our bodies and then converting those felt dynamics into lifeless language. For this reason, Sheets-Johnstone depicts language as "post-kinetic" (506).

Overview

I see so much within dance as a means to engender and facilitate some of the most valuable personal and interpersonal values, knowledges, and ontologies we experience as persons in a world that means so many different kinds of things. Dancing is a near universal talent of conscious persons, human or otherwise; its power of communication and expression is unbounded. For this reason, I move deeply into the epistemological capacities of dance for knowledge and cognition. What makes dance such an exemplar of the enactive model of cognition is that it is possibly one of the most outright and explicit modes of world-enacting. As a creative force, dancing brings to life worlds that can then be known, shared, and told by persons. Narrative dance establishes its own social ontology through an embodied process of telling stories, or storying, that *enacts* the world through meaning construction that make life more intelligible. I posit that dance not only permits of methods through which to express ourselves and tell our stories, it has within it various processes for which to ask the very questions we need to know to even be able to tell stories or share meanings. Dancing aligns with Johnson's account of worldmaking since dancing embodies the ideas of the world, and their dynamics within it. Epistemologies of dance generate multiple somatic, ontological paradigms in that they are a conception and an enactive dialogical praxis. Dancing as gesturing sits—well, dances—next to enactivism and enactive metaphors.

As Di Paolo et al. (2018) show, it is because we are inherently linguistic bodies that we have the means through which to advance our sensorimotor coupling with the environment to something as rich as the world in which we live. We enact our world through our interactions with others in embodied practices of sense-making. These enactivists take great measures to illustrate why and how we need to move beyond the cognitive linguistics of Lakoff and Johnson. Our embodied metaphors tell us much, but we have to understand our body itself as the origin, form, and content of meaning and language. Our bodies are not merely the vehicle for language, they language—period—regardless of whether we intend them to say something or not. They argue that linguistic bodies are made up of embodied utterances and the relations between those utterances; and our agency, which is a specifically linguistic agency, consists in orchestrating and ordering these embodied utterances, where utterances are recognized as acts that are embodied, material patterns enacted by sensorimotor bodies. Linguistic bodies and embodied utterances give rise to processes and practices of sense-making and interpretation in a shared world (2). Moreover, our bodies are always engaged in "critical participation" with other sensorimotor bodies by choosing and actively questioning and changing the frames of discourse. They say that an ethics-is-practice is realized in keeping ourselves open to our own unfinished embodied becoming and learning (3).

Part of what goes into the sense-making required for world enacting is questioning…if we take them seriously about bodies being capable of actively questioning—which I do. This recognition of our bodies as the full and unimpeded or unmediated—though not solitary—site of language is how we merge embodied metaphors and enactivism in a theory of embodied critical inquiry. What kind of world do we want to live in? What kind of values do we want to lay down for our interactions? What sorts of ontologies do we want to construct to make our world real to us? But beyond these existential questions, we have—in each of us—questions about our relationship to these larger enacting questions. What are our relationships with others? Do our communities contribute to our agency? Have we created a world in which we cannot thrive? Am I worth anything in this world? Am I beautiful? This understanding of linguist bodies as active questioners of the enacted world can best be seen through the context of dance since dance is an intentional mode of personal and interpersonal processual dialectics. As Robin Grove says, a dancer's language is body utterance; it is "…the body *being* uttered by a language is doesn't entirely know" (cited in McKechnie 2002, 159).[17]

In this book, I bring forth a framework for how to make sense of a mover's claim that movement expressions/ moving speech acts are more reliable utterances than verbal speech acts. I do so through the development of the embodied critical inquiry movement methodology. At the heart of my method is the foundational premise that intrapersonal and interpersonal movement dialogues can be manifested through somatic-based improvisational movement sequencing. An initial glaring question is whether I am doing something here that dancers have not considered before. There are two responses. First, what I provide here is not something that some choreographers don't know or don't know how to do. But what I do offer, in the theory, is an explication of why and how what they do in their studios is so entirely effective. Second, there are many movers, choreographers or no, who appreciate the intuitive validity of the potential of embodied critical inquiry but themselves do not quite know how to make sense of it or go about doing it. Therefore, I flesh out the complicated process of how one can even find embodied questions to respond to narratively. Third, I have not yet encountered another existing empirical methodology for choreographic and movement practices to guide an embodied critical inquiry. This empirical method is what I consider to be my most significant contribution to both enactivism and dance theory.

In Chap. 2, I highlight the role of creativity in the embodied mind account of cognition. The enactive account of embodied cognition rests on the supposition of the phenomenological body as a dynamic force in its situated contexts. Similarly, the account of creativity most closely associated with cognitive processes are dynamic factors related to one's phenomenological relationship to their context. Given that my account of embodied critical inquiry is grounded in the specific process of movement creation, rather than memorization, an understanding of cognition as embodied, creative, and dynamic is essential.

[17] For more general and related background work on this idea, see: Graham McFee, Graham. *Understanding Dance*. New York: Routledge Press, 1992. Randy Martin. *Critical Moves: Dance Studies in Theory and Politics*. Durham: Duke University Press, 1998.

In Chap. 3, I focus on the cognitive functions of affect as it relates to embodiment. Specifically, I show how affect lines up with understandings of intuition and kinesthetic empathy in a way that leads us to a conception of embodied intuition. This embodied affective intuition is imperative to understanding how an embodied critical inquiry is even possible. In the phenomenology of dance, affect is central because of its role in embodied cognitive processes, such as kinesthetic empathy, intuition, and body memory. To be able to tap into one's inner most sources of implicit procedural knowledge, one must be able to recognize and interpret the affective meanings of intuitive driving forces. One can connect the phenomenology of affect to the phenomenology of moving movement-knowings. To do so, I examine why and how consciousness is tied directly to affect to better understand the role of affect in intuition. I hold that if one can understand how affect is experienced in, and through, movement, they can better learn how to access and direct embodied knowledges.

In Chap. 4, I differentiate between improvisation and choreography using somatic practices as the dividing line. I include the signature features of each mode, as well as their sometimes-divergent purposes in the dance process. Most importantly, I draw out the differences between the two to separate them for my own methodological purpose; however, I should note that the lines between improv and choreography are often blurred or used concurrently within the creative processes. I argue that the somatic practices that are so closely tied to the creative process of improvisation are the cornerstone to an embodied critical inquiry method. By looking at the role of somatic practices in dance movement therapy, it becomes clear that soma is the life blood of our cognitive unconscious; and it is through somatic practices that we are able to access our embodied implicit knowledge. Specifically, our somatic practices bring us in touch with our embodied affective intuition so that we may direct our cognitive gaze inward. I argue that the use of somatic movement practices in improvisational movement practices makes embodied critical inquiry possible.

In Chap. 5, I flesh out the underlying theory behind an embodied critical inquiry method. In the first section, I provide a more detailed exposition of Sheets-Johnstone to explicate how and why the thinking-in-movement model is a discrete epistemology. This task is important because it is possible for an unarticulated, assumed epistemological framework to work against the explicit model of embodied cognition. Additionally, the specific kind of epistemological frame that grounds thinking-in-movement is important for explaining how, from a cognition perspective, dance can actually ***inquire into***, and solve, conceptual problems. In section two, I take a brief interlude to address embodied metaphors more thoroughly since these are active throughout the chapter. In section three, I take on the question of how movement and dance are an unmediated means through which to query. This section is the core theory of the method. While I try to avoid going into methodological exercises and practices, I do discuss how somatic work and embodied metaphors do the heavy lifting involved in embodied questioning.

In Chap. 6, I focus strictly on how the phenomenology of dance and an embody critical inquiry fit within an enactivist framework. More specifically, I look to various components of enactivist models to clarify how the various parts of an embodied critical inquiry are enactivist. In section one, I demonstrate how the phenomenology of dance counts as a naturalized phenomenology. In section two, I provide a thorough summation of Di Paolo, et al.'s theory of linguistic bodies, since I find this enactivist model to be best suited to helping understand how bodies can communicate all the way down. A rendering of bodies as fully communicative is indispensable for understanding how somatic exercises can operate as conduits to embodied knowledge, which can then be expressed onto the body through movement. In the final section, I revisit embodied metaphors. I switch to Katan-Schmid's version of embodied metaphors that she terms dancing metaphors to corroborate my contention that embodied metaphors are categorically compatible with enactivism.

In Chap. 7, I provide the movement method. This method is finalized from my perspective but maybe not from other's perspective. Therefore, readers my interpret my method as still under construction but it is not. It is, however, incredibly flexible. It purposefully allows for nearly unlimited adjustment and creative license. But the version I present, is the one that I determined was perfect for me as the way I understand movement, somatics, and facilitation. I first go through the somatic exercises that I relied upon, since these seemed to be the core of the method. I then present the steps to the method, following with a section of some clarifatory remarks. The final section includes the worksheets the movers used during the process. Most of the movement practices I utilized are movement practices developed by other movement and dance practitioners, who I cite. The movement practices I identify as centrally relevant to my method largely function as open conduit movement practices; I set them up in such a way as to use them to access and home in on explicit knowledges in the body. So instead of using the movement practices to let knowledges reveal themselves, I use specific movement practices that have a particular affective destination. Finally, in Chap. 8, I go over all of the results of my methodological experiments. I set out all of the initial questions and concerns my partners and I had at the outset, how we went through the method, how we resolved the initial questions, and in what ways were some parts more successful than others. I include commentary from my movement partners regarding their experience with the experiment.

For many, this project will feel like two very different projects: one academic, one practical. This is true for many reasons. I am substantiating the theoretical through the practical. This is a philosophical project that relies on its application in real movement for verification. Additionally, this is a form of public philosophy, and the method is the public philosophy. However, for philosophers to want to engage this framework as a mode of public philosophy, they must actually have a firm grasp on the whole of the philosophy behind the movement approach. But more simply than all of those defenses, the difference in feel comes from the fact that I am also a varied person, as many of us are. I am both a trained scholar and a trained circus monkey. I am both an academic and a mover. I am both a theorist and a lay person. And just like many people, I am not always as integrated in my personalities as I would like to be. It would be ideal if I were able to write the philosophical framework in as casual prose as I did the empirical method. But as it stands, I am not. So I hope readers will bear with me but only when it suits them and only as it suits their needs. There are theorists who will have little interest in the movement method and movers who will have little interest in the technical philosophy. But really, this just sounds like a regular get together at my house, and probably yours.

Bibliography

Albright, Ann Cooper. "Life Practices". In *Oxford Handbook of Improvisation in Dance.* Midgelow, Vida (ed). New York, 2019: 25–36.

Bergson, Henri. *The Creative Mind: An Introduction to Metaphysics.* Andison, Mabelle (trans.). New York: Dover Publications, Inc., 2007.

Busch, Thomas. "Existentialism: The New Philosophy". In *Merleau-Ponty: Key Concepts.* Diprose, Rosalyn and Jack Reynolds (eds.). Stockfield: Acumen Pub, 2008: 30–43.

Davies, Eden. *Beyond Dance: Laban's Legacy of Movement Analysis.* New York: Routledge Press, 2006.

Dewey, John. *Art as Experience.* New York: Berkley Publishing Group, 1931.

Dewey, John. *Experience and Nature*. New York: Dover Publications, Inc., 1958.

Di Paolo, Ezequiel, Elena Clare Cuffari, and Hanne De Jaegher. *Linguistic Bodies: The Continuity between Life and Language*. Cambridge: MIT Press, 2018.

Dreyfus, Hubert. "Merleau-Ponty and Recent Cognitive Science". In *Skillful Coping: Essays on the Phenomenology of Everyday Perception and Action*. Dreyfus, Hubert and Mark Wrathall (eds.). Oxford: Oxford University Press, 2014: 231–248.

Evans, Fred. "Chiasm and Flesh". In *Merleau-Ponty: Key Concepts*. Diprose, Rosalyn and Jack Reynolds (eds.). Stockfield: Acumen Pub, 2008: 184–193.

Fraleigh, Sondra. "Consciousness Matters". *Dance Research Journal* 32, no. 1 (2000): 54–62.

Fraleigh, Sondra Horton. *Dance and the Lived Body: A Descriptive Aesthetics*. Pittsburg: University of Pittsburg Press, 1987.

Fraleigh, Sondra Horton. "Somatic Movement Arts". In *Moving Consciously: Somatic Transformations through Dance, Yoga, and Touch*. Fraleigh, Sondra (ed). Urbana: University of Illinois Press, 2015b: 24–49.

Fraleigh, Sondra Horton. "Phenomenology and Lifeworld". In *Back to the Dance Itself: Phenomenologies of the Body in Performance*. Fraleigh, Sondra (ed.). Urbana: University of Illinois Press, 2018: 11–26.

Gallagher, Shaun. *Phenomenology*. New York: Palgrave Macmillan, 2010.

Gallagher, Shaun. *Enactivist Interventions: Rethinking the Mind*. Cambridge: Oxford University Press, 2017.

Gallagher and Dan Zahavi. *The Phenomenological Mind*. New York: Routledge Press, 2012.

Gallagher, Shaun and Robb Lindgren. "Enactive Metaphors: Learning Through Full-Body Engagement". *Educational Psychological Review* 27 (2015): 391–404.

Gallagher, Shaun and Zuzanna Rucińska. "Prospecting Performance: Rehearsal and the Nature of Imagination". *Synthese*, 2021: 1–19.

Gendlin, Eurgen. "The Wider Role of Bodily Sense in Thought and Language". In *Giving the Body Its Due*. Maxine Sheets-Johnstone (ed). Albany: SUNY Press, 1992: 192–207.

James, William. *Pragmatism and Other Writings*. New York: Penguin Books, 2000.

Johnson, Mark. *In the Body, In the Mind: The Bodily Basis of Meaning, Imagination, and Reason*. Chicago: University of Chicago Press, 1987.

Johnson, Mark. *Moral Imagination: Implications of Cognitive Science for Ethics*. Chicago: University of Chicago Press, 1993.

Johnson, Mark. *The Meaning of the Body: Aesthetics of Human Understanding*. University of Chicago Press, 2007.

Johnson, Mark. *Embodied Mind, Meaning, and Reason: How Our Bodies Give Rise to Understanding*. Chicago: University of Chicago Press, 2017.

Katan, Einav. *Embodied Philosophy in Dance: Gaga and Ohad Naharin's Movement Research*. London: Palgrave Macmillan, 2016.

Katan, Einav. “Dancing Metaphors; Creative Thinking within Bodily Movements”. *Proceedings of the European Society for Aesthetics* 9 (2017): 275–290.
Lakoff, George and Mark Johnson. *Metaphors We Live By*. University of Chicago Press, 1980.
Lakoff, George. *Women, Fire, and Dangerous Things*. University of Chicago Press, 1990.
Laban, Rudolf. *The Mastery of Movement*. Hampshire: Dance Books, Ltd., 1980.
Langer, Susanne. *Feeling and Form*. New York: Charles Scribner’s Sons, 1953.
Langer, Susanne. *Problems of Art: The Philosophical Lectures*. New York: Charles Scribner’s Sons, 1957.
Langer, Susanne. From *Feeling and Form*. In *What is Dance?*. Copeland, Roger and Marchall Cohen (eds.). New York: Oxford University Press, 1983: 28–47.
McKechnie, Shirley. “Movement as Metaphor: The construction of meaning in the Choreographic Art”. *Proceedings of the 7th International Conference on Music Perception and Cognition*, Sydney (2002): 157–160.
Merritt, Michele. “Thinking-*is*-Moving: Dance, Agency, and a Radically Enactive Mind”. *Phenomenology of Cognitive Science* 14 (2015): 95–110.
Morris, David. “Body”. In *Merleau-Ponty: Key Concepts*. Diprose, Rosalyn and Jack Reynolds (eds.). Stockfield: Acumen Pub, 2008: 111–120.
Noë, Alva. *Action in Perception*. Boston: MIT Press, 2006.
Reason, Matthew. “Affect and Experience”. In *Experiencing Liveness in Contemporary Performance: Interdisciplinary Perspectives*. Reason, Matthew and Anja Lindelof (eds.). New York: Routledge Press, 2016: 83–96.
Rucińska, Zuzanna and Kenneth Aggerholm. “Embodied and Enactive Creativity in Sports”. In *Handbook of Embodied Cognition and Sport Psychology*. Cappuccio, Massimiliano (ed.). Cambridge: MIT Press, 2019: 667–694.
Sellheim, Berndt. “Metaphor and Flesh—Poetic Necessity in Merleau-Ponty”. *Journal of the British Society for Phenomenology*, 41, no. 3 (2010): 261–273.
Sheets-Johnstone, Maxine. “Thinking in Movement”. *The Journal of Aesthetics and Art Criticism* 39, no. 4 (1981): 399–407.
Sheets-Johnstone, Maxine. “Phenomenology as a Way of Illuminating Dance”. In *Illuminating Dance: Philosophical Explorations*. Sheets-Johnstone, Maxine (ed.) London: Associated University Presses, 1984: 124–145.
Sheets-Johnstone, Maxine. *The Primacy of Movement*. Philadelphia: John Benjamins Publishing Co., 1999.
Sheets-Johnstone, Maxine. *The Corporeal Turn: An Interdisciplinary Reader*. Charlottesville: Imprint Academic, 2009.
Valéry, Paul. Philosophy of the Dance. In *What is Dance?*. Copeland, Roger and Marchall Cohen (eds.). New York: Oxford University Press, 1983: 55–65.
Weiss, Gail. “Ambiguity”. In *Merleau-Ponty: Key Concepts*. Diprose, Rosalyn and Jack Reynolds (eds.). Stockfield: Acumen Pub, 2008: 132–141.

Welch, Shay. *The Phenomenology of a Performative Knowledge System: Dancing with Native American Epistemology*. New York: Palgrave Macmillan 2019.

Whitehead, Alfred North. *Process and Reality*, Corrected Edition. Griffin, David and Donald Sherburne (eds.). New York: The Free Press, 1978.

Whittock, Trevor. "The Role of Metaphor in Dance". *The British Journal of Aesthetics* 32, no. 3 (1992): 242–249.

CHAPTER 2

Critical Inquiry: Creative Movement and Embodied Cognition

The notion of critical inquiry is not one that's easily defined because it is conceptualized distinctively across academic disciplines.[1] But more importantly, it is both contextually and culturally specific. Often, our critical practices are considered valid only if they take the form of logical deduction or induction. Many critical methods based on inference, especially in some non-Western methods, are considered unreliable and incapable of being substantiated with certainty. Further, Western practices of critical inquiry can be engaged in for its own sake. It is not necessary for inquirers to do anything with the knowledge or products they acquire or break down; it is the acquisition of knowledge that matters, not how or whether it is applied or shared. Contrariwise, in many Indigenous cultures, it is considered rude for the hearer to badger the speaker for further explanations than what the speaker determines they are prepared to share. Critical inquiry in these cultures and nations takes the form of active listening and critical inference through trial-and-error practices. Since practice and action-based praxis occupies the center of many

[1] The difference between critical thinking and critical inquiry is one merely of purpose. Critical thinking is inherently creative and involves the analysis, synthesis, analogizing, metaphorizing, etc. of ideas. Critical inquiry is one kind of critical thinking that is the creative and innovative approach to asking questions. All critical thinking is done in response to one kind of question or another. Critical inquiry is the critical thinking activity of pursuing and constructing the questions that the critical thinking activity of answering follows.

S. Welch, *Choreography as Embodied Critical Inquiry*,
https://doi.org/10.1007/978-3-030-93495-8_2

Indigenous epistemologies, critical inquiry is frequently performed inferentially since the purpose is to learn through experience. Moreover, what is valuable in such critical inquiry is not the acquisition of knowledge; the value is in its application. Unlike traditional epistemology and philosophy of mind, not to mention traditional cognitive science, Indigenous, Africana, and a lot of Western feminist philosophy and science, broadly construed, posit the body as the milieu of, and for, cognition-inquiry. They rightly attest that processes of the body that accrue knowledge are understood implicitly and intuitively. I aim to show throughout this project, it is possible to make these queries explicit through the utilization of creative movement.

In this chapter, I begin to pull together the strands of my argument that an embodied critical inquiry is possible, valuable, and advantageous through creative movement, such as (or especially through) dance. First, I offer a succinct conception of critical inquiry. I purposefully sketch a general model of critical inquiry to ensure that it does not preclude any of the assorted modes of inquiry that are either outside of philosophy or outside of the Western canon; but my understandings of, and commitments to, critical inquiry practices align with the dictates of the aforementioned feminist and non-Western praxes and are strongly influenced by them. What is important here is not so much how critical inquiry is demarcated but, instead, whether and how crucial the capacities for creativity and curiosity are to our "critical" investigative work. Second, because my framework is situated within a dance context, and because choreography construction is a creative process, I put forward a survey of recent literature at the intersection of dance, cognition, and critical inquiry to indicate how dancer-scholars have merged them thus far. By looking at the established research on the interconnections between these areas, one can envisage the openings available for expanding on the subject. In section three, I bring together the critical aspects of creative movement and the notion of curiosity to illuminate how creative forms of movement are instantiations of active embodied practices of a curiosity-in-motion or "curiousing".

Critical Inquiry and Creativity

Critical inquiry broadly encompasses methodical practices of knowledge-seeking-investigation. Inquiry can be done for many reasons, in any number of fashions. One might ask questions or attempt to interpret body language. One could follow a story to its conclusion, and equally so, a path. One can take things apart and look inside and maybe even try to put them back together to see if they can make the parts fit in the same, or different, way. But inquiry simpliciter is not necessarily critical in nature.

One aspect of critical inquiry is a "methodological" structure. By methodical, I mean structured and purposeful—one needs a plan, loosely speaking. Inquiry cannot be critical when it's willy nilly. Nay nay, it's not even critical if it has a purpose but that purpose is just to drive people bonkers. For example, women sometimes tap into their bellies to question whether someone is trustworthy to help them determine which actions they ought to take. Many women have their own way of going about testing their intuitions and they usually do so for specific kinds of reasons in specific kinds of contexts. Wine tasters have particular ways of attending to their senses when attempting to pinpoint a specific ingredient in the elixir. Others might have their own little process for gathering information to respond to a problem that absolutely *must* involve post-it notes. Any person can utilize any or all methods of inquiry for various situations. But what is important is that they intend to go about the inquiry in active detective mode.

Relatedly, one trait involved in detective work is having a target to detect; that is, one must seek to learn something, even if they do not know what it is they want to learn and do not know for sure what it is that they seek to find; it is not uncommon for people to Eat, Pray, Love to hunt down the meaning of life. One can even engage in inquiry of a critical nature if they don't particularly want to find what they are looking for, including the *actual* meaning of life. Our bellies tell us many things we do not wish to know but yet we ask anyway. Often, even though Western theorists claim to be seeking knowledge in its own right, this isn't particularly accurate. For even if they do not aim to make use of the knowledge gained, they do seek it to solve a problem, no matter how imaginary its origins or futile to society. In this case, the scholar learns something and applies it to something, such as a gap in their own knowledge; it is just that they don't particularly care to share it with anyone else so that it could be useful.

But another necessary feature of critical inquiry is, at the very least, a touch of creativity. One's creativity can be recognized in the distinctive way that they assess the problem at hand. For if they were not approaching a situation uniquely, then there would already be a solution available to them on Google. A person may recount their experience to a friend who has shared a similar experience in order to receive advice on how to proceed. What marks that spark of creativity is their ability to line up their experience to their friend's experience so that they can attempt to figure out how far, and to what degree, their friend's solution is suitable to their own situation. Those who lack this capacity, or who choose not to make use of it at the time, are not engaging in a critical form of inquiry; they are seeking a prefabricated plan of action to mimic. Figuring out whether one's attempts at inquiry are critical is not particularly hard to do. Critical intentions are typically starkly contrasted with random forms of questioning. What is hard about identifying or differentiating between similar presentations of querying behavior is pinpointing the creativity in it, wherever it may be.

In literature on creativity, the supposition that it is already related to a/ the/ some process of critical thinking qua critical inquiry is a given.[2] This is because creativity also has purpose and method. Even the bullshit some people spout off at after-parties to sound creative is actually creative; they had to figure out what to say to ensure others perceive them as creative. So while the content of the utterance is bullocks, the process to get there was creative, as annoying as it may be to admit. Indeed, one must be clever to portray one who is clever. This is also why some of our best cheaters and scammers are actually quite creative masterminds and not the lazy bums they are made out to be.

Creativity is similarly a slippery concept and so hard to pin down. Some people think creativity is culturally and contextually dependent, some think it must be recognized as creative by the public, some think creativity is completely subjective, and some think it is relative to one's creative domain. Some agreed upon criteria for a creative product is that it is novel, original, and atypical. Some believe that anyone can be creative, but others think there is something unique to creative individuals that ordinary folx lack. Creative personality traits include traits of individuals who are largely introverted or private, highly open-minded, playful, patient, risky, driven, ambitious, self-aware, and highly independent (Plucker et al. 2019, 50). Because creative behavior pushes the boundaries of social, artistic, academic, technological, or aesthetic norms, these traits of creative individuals are often perceived as hostile, complex, and impulsive (ibid.).

[2] See also: James Kaufman and Robert Sternberg (eds.). *The Cambridge Handbook of Creativity*. Cambridge: Cambridge University Press, 2019. Margaret Boden (ed.). Dimensions of Creativity. Cambridge: MIT Press, 1994.

While folx disagree about the subjective nature of creativity and the conditions for it, there is a considerable consensus among theorists regarding the cognitive conditions involved in creative behavior. These more accepted traits are not necessarily tied to one's personality, and they are also not necessarily tied to one's intelligence—though there can be some correlation (Benedek and Jauk 2019, 203). Additionally, cognitive traits can be learned and so go against the assumption that creativity is a natural personal talent. One of the commonly accepted signifiers of creativity is one's ability to put forward multiple responses to an open-ended prompt. The ability to do this, however, relies on one's ability and capacity for higher cognition skill execution.[3] Psychologists often use the Cognitive Research Test (CoRT) to identify common underlying skills requisite for producing creative behavior and products (Sternberg 2019). These cognitive skills include: flexibility, imagination, lateral thinking, divergent thinking, generation, synthesis, provocation, affect, and analytical thinking. One of the most recognizable models for assessing cognitive creativity is the Geneplore model, which focuses on the generative and exploratory phases of creative activities (Finke et al. 1996); the creative process is usually a cycle between these two phases as the person goes through and assesses various responses and possibilities to prompts. The generative process includes: retrieval, association, synthesis, transformation, analogical transfer, and categorical reduction (20). Persons moving through the creative process execute these cognitive skills and apply or manipulate them through "preinventive structures", such as visual patterns, object forms, mental blending, category exemplars, mental models, and verbal combinations (ibid.). Regarding creative behavior, Robert Stenberg asserts that "[i]t is far more important for students to figure out *what questions to ask—and how to ask those questions*—than to learn the answers to the question" (93, italics mine).

Both critical inquiry and creativity are disorderly by their very nature. They require that people dig and pull things apart, incorrectly piecing things together until they can do so correctly—or they just so happen to like what they originally put together even though it doesn't actually fit that way according to standard expectations. Some people are drawn to

[3] These cognitive tasks do not require a specific kind of cognitive normativity because creativity does not require all of these conditions. One can be creative solely through imagination, or affect, or what not. But also, the other higher cognitive skills in the list are also relatively available to cognitive and neuro diverse people.

this because they desire to pull things apart and put them back together again. Others find these practices unmanageable. And there are folx who aren't into this game but like to revel in the chaos of deconstruction for its own sake. This doesn't mean that these people are not creative or that they do not inquire. It just means that some people may find other ways in which to be creative or critical. Deconstruction can be said to be a form of critical inquiry since critical inquiry does not entail building up again where what you tore down once stood. But as a general practice, engaging in negative inquiry, at the very least, calls for positive response. In this sense, one who loves the demolition stage of critical inquiry for itself is one who may not be interested in what comes after, such that they become more of a provocateur-gadfly than a provocative thinker.

Curiosity is what drives people to look under the rug to see whether there's a trap door underneath. It is as important to critical inquiry as is creativity; Philipa Rothfield astutely remarks that not knowing how things will turn out is a kind of opening in itself (Rothfield 2019, 99). Another way to conceive of curiosity is to bank it on a notion of wonder. Iris Marion Young (1997) explains that for one to take a querying stance amenable to a perspective of wonder, they must orient themselves into an asymmetrical relation to others and the world. For one to avow a willingness to listen but hold on to an assumption of sameness in perspective would undermine the importance of asking questions. Instead, she argues, individuals should pursue an asymmetrical reciprocity in perspectives that prioritizes difference and plurality. Albright intimates a similar idea when she claims that improvisation is a philosophy of life that adheres to a willingness to relate to other's experiences to simply find out about them (Albright 2019, 25). Here we can see the overlap between the conditions for creativity, critical inquiry, and curiosity. Wonder is what we experience when we cannot fit an idea or experience into our accustomed lived configuration. Individual creativity and curiosity are fostered through interaction and engaged, dialectical, exploratory learning practices that are question-centered for the purpose of encouraging reflection, understanding, and demonstrating respect. Imagination and wonder are key to generating understanding through critical inquiry.[4]

[4] See also: Thomas Ward and Yuliya Kolomyts. "Creative Cognition". In *The Cambridge Handbook of Creativity*. Kaufman, James and Robert Sternberg (eds.). Cambridge: Cambridge University Press, 2019: 175–199.

Gallagher et al. (2015) connect the notion of wonder to that of awe. Creativity is incapable of driving us to innovation if we cannot be amazed and overwhelmed by that which we experience, and even more so, that which we do not but could. While wonder may be (mistakenly) conceptualized as an ethereal reflective perspective, awe is undoubtedly affective and a fully embodied cognitive experience. They define awe as a direct and initial experience or feeling when faced with something amazing, incomprehensible, or sublime (6). Consequently, the experiences of awe and wonder drive our curiosity towards the tasks of tearing down/tearing apart and building up. They rightly attest that:

> Ineffability, actually, may occur in either ontological or cognitive wonder, but for different reasons. If there is the perception of a larger order or reality for which there is not ready language, there might be gaps. On the other hand, if there is a concept that has no relation to other concepts, there might also be a lack of language to express this unique thing. In either case, stumbling and searching for words may be an indication that there is an experience worth attending to. In some cases, wonder may exist not in the concepts but in the gaps between concepts that are inadequate to the phenomenon. (122)

To be fair, the experience of wonder is more recurrent and relatable among people with shared meanings and experiences; and it is more likely to lead one to, and through, the inquiry process. Awe, on the other hand, which resembles something more like a mystical encounter, is more likely to bring one to their knees with the overwhelming weight of incomprehensibility or incommensurability. Only after one makes peace with the greatness of awe can their more manageable wonder put their critical queries into motion.

Underlying the phenomenological conception of the experience of wonder is partially what is termed the "feelings-as-information" model (Gotlieb et al. 2019).[5] This model undergirds wonder insofar as creativity relies on our imagination and our imagination largely relies on our ability to experience empathy and take on others' perspectives. Because creativity depends on our ability to conceive of the novel, cognition makes use of

[5] For the specific role of moods, rather than affect, see: Matthijs Baas. "In the Mood for Creativity". In *The Cambridge Handbook of Creativity*. Kaufman, James and Robert Sternberg (eds.). Cambridge: Cambridge University Press, 2019: 257–272.

pretend play with possible affordances to imagine something not already extant. On this model, the failure to be creative can result from an under-developed recognition of, and experience with, differently situated others in a differently contextualized world. Imagination and make-believe are not individual/personal cognitive practices. These practices require participation in, and an understanding of, a social world full of diverse others. This is because our experience of others provides more associations for us to draw from when attempting to enact metaphors. The depth and breadth of our social imaginary lends itself to our particular, unique imagination and helps us connect complex ideas (714). Further, our ability to take on other's perspectives and to experience wonder and awe in doing so, is tied to our emotions since emotion is their vehicle—emotion fuels creativity (Ivcevic and Hoffman 2019, 273). Positive emotions generally, and sympathy and empathy specifically, contribute to originality because they make one sensitive to peripheral and social cues and thus gives us access to a wider range of imaginative fodder needed for critical thinking (281).

All of this goes to show that creativity and curiosity, and thus, critical inquiry are, or involve, embodied "higher" cognitive processes and skills. But affect and social interaction is not the full extent of the embodiment of critical inquiry. Our sensorimotor contingencies also do some heavy lifting in the generative and exploratory phases of our querying. Most obviously, our sensorimotor contingencies shape the specific affordances that we can recognize in our interactions with others for the purpose of make-believe later on. But less visibly, our sensory perception is activated during our imaginative processes. Gallagher tells us that "...memory or imagination involves a (re-) activated presentational activity that evokes or brings to presence something that is absent..." (2017, 188). Imagination is a "subjective simulating" of our perceptual experience, which is necessarily embodied. Moreover, while traditional cognitivists posit the imagination as a higher cognitive function that is fed on mental representations, enactivism points out that the imagination is a doing—an action—since it is actually fed from our active sensorimotor engagement with possible affordances (191, 193). The recognition of sensorimotor and affective components of imagination helps one make intelligible the more abstract claim that critical inquiry is an active, movement-based embodied doing rather than a passive, armchair marathon.

Creative Cognition in Dance

Over the past few decades, there has been an uptick in the various ways in which dance is addressed in relation to critical thinking. One of the more common direct links between dance research and critical thinking is in educational processes. Dancer-scholars and education theorists point to the ability of dance to enhance our cognitive capacities.[6] Thus, dance classes can augment children's ability to successfully engage in problem-solving. Dance has been recognized as obviously contributing to one's sensorimotor awareness, which then transfers over into spatial and analogical thinking. But it also broadly impacts an individual's ability to problem-solve through choreographic exploration that encourages one to stray from the path, as it were, in order to make one anew. Dance teachers also expose students to various kinds of critical thinking and problem-solving activities to work in the other direction.

Training in abstract thinking skills bolsters dancers' creative choreographing since it helps them to imagine movements disconnecting and reconnecting, especially by conceiving of open spaces as new affordances where none were previously seen. Karen Barbour (2011) gives an excellent account of how dancing is a critical process of discovery, which is a specific mode of embodied research. Even on its own, choreographing—when taken up purposefully—allows dancers to uncover, examine, explore, and then express their relationships to the world. She describes one dancer's experience in choreographing as itself being the source of "liberating" knowledge (93). She argues that choreography, as an embodied form of coming to know, can instantiate as a sort of "literature review" of their

[6] See: Karen Barbour. "Embodied Ways of Knowing: Revisiting Feminist Epistemology". In *The Palgrave Handbook of Feminism and Sport, Leisure, and Physical Education*. Louise Mansfield et al. (eds). New York: Palgrave Macmillan, 2018: 209–226. Michael Gard. "Movement, Art, and Culture: Problem-Solving and Critical Inquiry in Dance". In *Critical Inquiry and Problem-Solving in Physical Education*. Wright, Jan, Doune McDonald, and Lisette Burrows (eds.). New York: Routledge Press 2004. Ann Thomas Moffet. Dance as Inquiry: Critical Thinking in Dance Education. Thesis. Department of Dance. Graduate School of the University of Oregon, 2010; Anna Pakes. "Original Embodied Knowledge: The Epistemology of the New in Dance Practice as Research". *Research in Dance Education*, 4, no. 2 (2003): 127–149; Catherine Stevens. "Trans-disciplinary Approaches to Research into Creation, Performance, and Appreciation of Contemporary Dance". In *Thinking in Four Dimensions: Creativity and Cognition in Contemporary Dance*. Robin Grove, Catherin Stevens, and Shirley McKechnie (eds.). Melbourne: Melbourne University Press, 2005: 154–168.

own embodied experiences and knowledge in dancers' research (97). In her work, she offers a set of guidelines for doing this critical form of embodied research. These include: (1) acknowledge that everything is possible and potentially relevant—including movement, intuition, and lived experience, (2) dance researchers should engage in art-specific learning, (3) they must *play* through actively experimenting and improvising with new questions and challenges, (4) they should partake in, and be attentive to, practices of trial and error to re-create and reflect on patterns, relationships, and themes, and (5) they should be flexible with approaches.

But there are other areas of research that connects dance with critical thinking more indirectly. Work on kinesthesia draws heavily on dance as a site and analogy for analysis. Susan Stinson argues that theory and one's ability to theorize evolves from kinesthesia since thinking is an active activity. On her account, thought itself is experienced throughout the body (Stinson 1995, 46); she explains that thinking, reflecting, and analyzing are not simply in the mind. Further, she argues that we cannot be thought to possess knowledge without having secured that knowledge deep within the body (ibid.). Similarly, Jane Carr posits that the dancer's experience of the dance is a significant component in theorizing about its overall meaning and significance because they have an "inbuilt self-awareness" that thrives through dance training. By incorporating the dancer's perspective qua self-awareness into the understanding of the dance, theorists can move between different sensory modalities to better depict its meaning (Carr 2013, 73). This distinctive kinesthetic self-awareness leads theorists to consider dance as a paradigmatic mode of perception in expert action.

Many theorists/coaches/athletes argue that in-the-moment skill execution is "thoughtless" insofar as it occurs automatically from repeated and consistent habituation—in effect, it is effortless. This perceived effortlessness is often referred to as the just-do-it principle. However, as Barbara Montero rightly asserts, expert skill execution is intensely aware and thought-driven. The cognition-in -action principle within embodied cognition theory reveals how expert performance requires conscious deliberative processing. She specifically draws on dance to clarify this point by drawing a distinction between an expert appearing to be effortless and actually acting without conscious effort. She claims that "…one reason to think that experts or at least expert dancers, focus on their movements rather than just doing them, is that they experience their movements and positions as beautiful, graceful, powerful, precise" (Montero 2016, 192).

It is the body that is explicitly engaging all of the conscious processing and that intense conscious engagement and embodied deliberation is necessary to be so completely immersed in action. If there were no conscious processing and explicit cognitive effort, many landings would fail. Montero's argument is another example of how dance is typically connected to critical thinking because she uses this argument to articulate the actuality of creative deliberation in action.

But the way in which we can begin to think about movement and dancing as a mode of embodied critical *inquiry*, rather than as a mode of embodied critical *thinking*, is to understand how movement is connected to creative cognition. As it is, movements, such as gesture, are already strongly tied to creative cognition. Gestures are the languaging mode of the body. Gestures are naturally and innately communicative quite independently of verbal language; but they are nearly impossible to occlude from verbal linguistic communications. Gestures are spontaneous in the moment, but individuals develop, over time, what appears to be a person style of gesturing. Much like most people have a particular dance style or signature (even those wedding-only dancers), every person has a distinguishing gesture signature. Typically, we "make decisions" about which gestures to impart nonconsciously and eventually our gesture creations become part of our body habitus that still varies in contexts—people develop gestures that are enacted in professional contexts quite differently than in personal relationships—though it will still have the same root "word". Because gestures represent thought in action; they are to be understood as thought itself insofar as thought is recognized as a form of inner discourse, which subsequently plays out creatively through gestures. For example, gesture code-switching is a thing. Gestures are not merely reactionary reflexive motor actions that signify the highs or lows of the story or the importance of the communicative transmission. Most importantly, gesture lets us watch the thought of others as they manifest (McNeil 1992, 132). Similarly, gestures, much like body language, can and will betray a speaker's verbal communications; if one uses unexpected or atypical gestures with a partner; the partner can decipher the inconsistency in what is being communicated. Because gestures represent actual thoughts qua embodiment, they can expose the thoughts of the speaker (ibid.).

When explaining the embodied-cognitive-linguistic nature of gesture, Gallagher informs us that the meaning-movement involved is necessarily an expressive movement. He argues:

> The body materializes language by means of movement that is already expressive. Like language itself, expressive movement must be both natural and conventional, shaped by innate mechanisms that allow for the onset of communicative behavior, and by cultural determinants that define different languages and linguistic practices. Although speech and gesture depend on movement as a necessary condition, they nonetheless *transcend* motility and move us into a semantic space that is also a pragmatic, intersubjective, intercorporeal space. Meaning overflows the movement that is involved in gesture and speech, and, just so far, language transcends embodiment at the same time that it depends on it. (Gallagher 2006, 126, 127)

Similarly, Robin Grove explains that:

> …part of the reward of watching a dance is not only to receive communication of affect and meaning from the dancers, but also to experience that affect and meaning as expressed by our own bodies. (Grove 2005, 44)

And Cynthia Bull cites a dancer's explanation of this cognitive communication as having a social component as well:

> …my *sensible* dance experience includes and implies *intelligible* choreographic and social meanings…My dancing also stirs very personal associations and images within me as I move, while my audience creates many different narrative and imagistic interpretations of my dancing. (Bull 1997, 269)

This recognition of dancing body gestures and their cognitive communicative powers are now being centered as the direct object of research. Art, generally, organizes our cognitive worlds of imagination to open more affordances; art, as such, offers venues for the communication of the distinctive experiences of our potentialities (Grove and McKechnie 2005, 4; Tweney 2005). Dance, specifically, produces and contributes to our consciousness because dancers perform on, and of, themselves directly. This movement product of consciousness is effective precisely because thoughts are not held, but rather are experienced. The newly recognized pathway between dance and cognition as a form of creative cognitive communication, at least equal to that of gesturing, motivates much of the research of dancers and dance lovers in cognitive science and neuroscience.[7]

[7] See also: Susan Leigh Foster. "Making a Dance/ Researching through Movement". In *Mapping Landscapes for Performance as Research: Scholarly Acts and Creative Cartographies.* Shannon Rose Riley and Lynette Hunter (eds.). London: Palgrave Macmillan, 2009: 91–98;

McKechnie and Stevens identify the cognitive nature of choreographing as choreographic cognition, which is comprised of perceptual, cognitive, and emotional processes as a dynamical system (McKechnie and Stevens 2009, 38; Stevens 2005b). In their research, as somewhat paralleled to Katan-Schmid's and my own on metaphors and mirror neurons, they find that the novelty of choreographic creativity comes from metaphorical thinking (39). They demonstrate how complex dance vocabularies counter the representationalist view of cognition vis-à-vis memory as a collection of propositional knowledge; rather, they show that dancing is declarative through procedural knowledge because the language of the dancer is an "utterance of the body being uttered by a language it doesn't entirely know" (45). Like gesture, dancing is a way of languaging. Through their study on fMRIs, they reveal that professional choreographers analyze movement and exact the symbolic units that activate the semantic networks associated with particular gestures and actions (46). "Verbal language strings together sequences of words, and dance strings together sequences of movement to make phrases and sentences" (Hanna 2015, 27). Likewise, Stevens, et al. argue that cognition cannot be strictly representative since dancing helps us see that it makes no sense to conceive of an active dynamical system of sensory information that imbues dancing as representationalist (Stevens et al. 2003b, 318–319). Following these findings of the relationships between cognition and dance, Rosaleen McCarthy et al. (2006) investigate dancers' cognitive skill enaction during movement sequences whilst simultaneously performing sensory and proprioceptive tasks that align with the same cognitive resources as those involved in dancing.[8] And in a corresponding project, Judith Hanna argues that our brains actually choreograph our sensory experiences, thoughts, and actions (Hanna 2015, 1). She claims that we think through our brain-body as a way to process our physical experiences that are culturally constructed. On her account, the cognitive activities deployed in the motor functions of dance trigger our

Catherine Hale. "The Science of Making Dances: Report on Choreography and Cognition Event at Winchester Festival". *Dance Gazette* 2 (2004): 16–19; Shirley McKechnie. "Choreography as Research". In *Creative Investigations: Redefining Research in the Arts and Humanities.* Stoljar, Margaret (ed.). Symposium of the Australian Academy of the Humanities, 1995.

[8] See also: Ivar Hagendoorn. "Dance Perception and the Brain". In *Thinking in Four Dimensions: Creativity and Cognition in Contemporary Dance.* Robin Grove, Catherin Stevens, and Shirley McKechnie (eds.). Melbourne: Melbourne University Press, 2005: 137–148.

conceptual processing (17), thus making cognition—at least, during the *doing* of dance—creative and embodied.

Relatedly, Dunagan et al. (2019) discuss the significance that dancers' and students' recognition of criticality and coordination at deep cognitive and sensorimotor levels reveals the puzzle-solving nature of interactive movement and choreography itself. Choreography specifically illuminates the multitudinous ways in which creativity is a function of movement synthesis, connection, and associations in the face of these moving puzzles (558). The puzzles of choreographic movement come not only from the rules of technique and ability but also from the dynamics of intersecting and overlapping bodies. They argue that critical thinking goes into, and comes out of, the dancing process. Critical thinking is inherent to the creative movement itself in terms of a distinctively distributed creative cognition across the group members. Dance, they say, is a puzzle to be solved—through creative movement and creative bodying. In short, dancing *is* critical thinking because it requires active kinetic decision-making *in* motion.

Creative Movement and Curiosing

But there are a few aspects about movement in terms of meaning-making that need to be clarified before one can be said to be a proper mode of critical inquiry. For as some embodied theorists suggest, movement itself may not be a site of the higher form of cognitive processes, even if those processes rely on sensorimotor contingencies and capabilities resultant from previous movement. The first step is to understand the distinctive nature of specifically creative movement. This conception of creative movement is going to be aligned more so with the understanding of meaning-making movement as expressive, such as in the case of gestures, than it is with the more expansive but general conception of movement and orientation as the generator of body schemas, such as being upright. Because expression is personalized, stylistic, and context or relationship dependent, that puts expressive movement at the center of creative movement. But if we want to designate it as "properly" creative *cognitive* movement, we must look to see if expressive movement, including but not limited to dance, satisfies some of the conditions for creative cognition set out earlier.

Even if we limit the assessment of choreography to the most general socially shared understanding of choreographing, there are a number of cognitive skills that are at the fore of the process, such as mental flexibility, imagination, provocation, and affect. These skills constitute the minimal ground for the creative cognitive process of building, changing, shaping, and piecing together movement vocabularies. These skills, then, easily fall under activities that require the manipulation and application of the preinventive structures—visual patterns, object forms, mental blending, mental models, and bodied-verbal combination—because the manipulations of these preinventive structures are the structures through which the basic creative cognitive skills are enacted through movement creation and ordering. And obviously choreographing cycles through the Geneplore model's two phases of generation and exploration. This is the create-and-revise process that drives the creative-cognitive-affective push needed to motivate an artist's long-term investment.

At first glance, when looking at the cognitive skills comprising the generation phase— retrieval, association, synthesis, transformation, analogical transfer, and categorical reduction—one might not immediately jump to creative dance as a cognitive exemplar. But this is to underestimate or overlook the ways in which the broadly accepted skills involved in choreographing, like imagination and mental flexibility, are the grounds for, and impetus of, generation and exploration. Cognitive retrieval, association, synthesis, etc. are embodied just insofar as the knowledge of movement is already in the bones. Certainly, choreographers can execute these cognitive skills when sitting on the floor but even in doing this, they are reaching inside their body to their embodied knowledge to determine which sensorimotor affordances are available to them. The imagination of choreographers, when conceiving of movement, shape, and dynamic possibilities is almost entirely kinesthetic; they are in no way fooling around with mental representations of body parts as if they were abstract puzzle pieces. The kinesthetic enactment engaged in by the artist during generation is a composite of these other higher cognitive skills. The embodied moving puzzle pieces of dynamics are retrieved, not from some storage bank in the brain, but rather, from within the experienced and experiential body. From here, one can see that the other higher cognitive skills signifying creative cognitive processes, such as lateral thinking, divergent thinking, generation, synthesis, analytical thinking, synthesis, and analogical reasoning, are activated and exercised.

Given the relationship between creativity and curiosity, it is important to look at how movement itself can be a mode of curiosity rather than merely a phenomenological activity driven by "mental/representation" curiosity. The creative cognitive skills are operations of formulation, deconstruction, and arrangement. When I posit that artists execute these skills in the generative and exploratory processes, I point to the artist's search for new affordances. From this perspective, it would appear to be the case that the movement exploration is driven by curiosity. But again, this assumption is based on a misunderstanding on the embodied process. I claim that the choreographing process is motivated by curiosity, but I would also insist that the movement is also, itself, curiosity, not merely an expression of curiosity—it is a curiousing, of sorts. The reason why I make this claim is because of how the body seeks out affordances in movement. The body does not move because the dancer *tells* it to. The body seeks affectively and kinesthetically. These embodied modalities are what drives the movement, and the movement seeks the affordances.

This articulation of movement-as-curiosity may be more than a little rough around the edges for a non-mover. But many movers can make sense of what it means for the moving body to be curious. For example, when I am doing aerial rope exploration, I do not tell my body to go hither and yonder. I do not tell myself to construct this wrap and then tell myself to poke a leg here or poke an elbow there. My body does this. My body wants to know. My body wants to find the affordances, where affordances are just actual or potential openings for engagement. When I am in proper exploration mode, I tap into myself and then I tap into the rope. We refer to this as having a connection with the apparatus. This is quite similar to when one has connected to their motorcycle. The two become one; the apparatus is, as some would say, the "extended mind". When I am bonded with the rope, my body does the poking and sniffing around; but again, I am my body. I am a minded-body-subject. The ways in which my limbs twist and turn and poke *is* my curiousing. My movement is curiously and critically inquiring as to what it should do, where it should go, how it can get there, and what it will find. When we understand movement-as-curiosity, we can recognize that the cognitive activity of being curious in and about movement affordances does not precede the movement; it resides in the movement. Curiosity is enacted in movement; movement enacts curiosity; curiosity moves; movement curiouses.

Another aspect of movement-as-curiosity that is useful to understanding the role of creative movement in embodied inquiry is its ability to be a mode of wonder and awe. Wonder is what we experience when we've opened ourselves up to a plurality of perspectives that aren't already constitutive of our understanding of lived experience. Our wonderous engagement with diverse perspectives is derived from, and contributes to, our dialectical exploratory learning practices. The same holds true for movement; as Sheets-Johnstone says, dancing allows one to wonder about the world directly through movement since thought itself is kinetic. When we move-curious, we are actively engaging in embodied dialectics. When I experiment with my rope, I listen to what it offers me. I come to the rope already with a preconceived conception of its potentialities in light of my own. But when I am exploring, I am admitting that my preconceptions about the affordances of the rope, based on the limitation I alone place on it due to my particular movement and wrap vocabulary, are incomplete and rife with…for lack of a better word…stereotypes about what it is and what it can do. If I cannot walk up to my rope with an attitude of kinetic wonder, I will never be able to explore, much less create. Wonder-movement makes movement-as-curiosity possible. Awe, on the other hand, is handed to me when my body finds openings that I could not possibly have conceived as available. Earlier I explained awe as an overwhelming affective embodied experience that makes one feel as if they are brought to their knees. In movement creation, this analogy becomes less figurative and more literal. In other words, awe is very rare and it is so because usually this affect emerges from a movement that one "ought not" have been able to make. There are often many moments in aerial when one's curiousing gets away from them and they move through a hole that is there for a reason…there's nothing holding it together. But every blue moon, instead of falling to greater and lesser degrees of injury, one will fall into a whole new wrap that they could not have expected to come together. One of the main ways in which wholly new and original movement vocabularies come into existence is by accident. But most frequently these accidents are also close calls. And in these situations, one assuredly feels awed by the experience of discovery.

Central to the understanding of movement-as-curiosity is a parallel conception of movement-as-problem solving. These two concepts combined ground the method for using creative movement as a directed process of critical inquiry. Surprisingly…or actually, not surprisingly at all, embodied metaphors pop up here to help explicate how movement can go about problem solving. But a few obvious questions arise when we consider embodied metaphors in their traditional articulation. For example, if embodied metaphors are often taken to be conceptual (usually in functionalist accounts of embodied cognition), how is it possible for them to be creative outside of verbal linguistic extensions? Further, it would appear that if embodied metaphors, even if they are fully enacted and not purely conceptual, have a determined meaning. What I mean here is that when we engage embodied metaphors linguistically, either through verbal or bodied languaging, embodied metaphors just are—or so it would seem. They exist as a function of our body schemas: their shapes, their orientations, their spatial relationality, and their historically dependent role in cultural linguistic expression.

While I take up embodied metaphors later in the analysis of embodied critical inquiry, I want to highlight the important work by Leung et al. in their article "Embodied Metaphors and Creative 'Acts'" (2012). In this article, they specifically address how embodying metaphors is a specific mode of problem solving, which facilitates arguments about embodied metaphors playing a large part in choreography as critical inquiry. According to traditional embodied metaphors accounts, the associations induced by the body-mind depicted in our metaphors are not "metaphorical". And according to these theorists, movement and moving lowers mental barriers. They argue that embodying metaphors for the purpose of creativity enhances generation because moving facilitates our psychological processes for creative problem solving (503). In line with what I have argued before, they point to the ability of embodied metaphors to tap into implicit embodied knowledge. They hold that by tapping into this embodied knowledge and embodied intuition, the movement itself is a process of problem solving because the embodied metaphors activate the cognitive processes that can overcome mental "fixedness" to forge new connections between ideas (507). More specifically, they argue that:

> ...embodiment can potentially enlarge—not just activate—repertoires of knowledge by triggering cognitive processes that are conducive to generating creative solutions. In other words, body-mind linkages influence not only processes of knowledge activation, but also processes of knowledge generation. Embodying metaphors for creativity appears to help ignite the engine of creativity. (508)

This claim harkens back to my earlier survey of how movement is viewed as a source of creativity that also has the ability to enhance cognitive skills for critical thinking. What is unique about their account, and especially beneficial to the issues under consideration here, is their specific focus on the ability of embodied metaphors themselves to do this work. As I just mentioned, one might assume that embodied metaphors are a given, almost to the extent that they are fixed, and thus are not capable of being purposefully engaged and utilized for creative embodied cognition modes of critical exploration. But their explanation of how embodied metaphors facilitate cognitive processes for problem-solving in a way that rises above fixed ingrained presuppositions and dispositions, opens the door for us to acknowledge their capacity to be exercised in critical inquiry practices. Embodied metaphors can link movement, creativity, curiosity, inquiry, and problem-solving.

There is a variety of ways in which movement makes meaning and is linguistic, expressive, constitutive, structuring, and processual in relation to our phenomenological experience and understanding of the world and ourselves in it. If movement is able to do so much and say so much for us—even if, and when, we intend to say nothing—the supposition that movement itself can be both the subject and object, the vehicle and the form, of critical inquiry should be quite sensical. If movement gives us the meaning, should it not also find us the questions that do so?

Bibliography

Albright, Ann Cooper. "Life Practices". In *Oxford Handbook of Improvisation in Dance*. Midgelow, Vida (ed). New York: Oxford University Press, 2019: 25–36.

Barbour, Karen. *Dancing Across the Page: Narrative and Embodied Ways of Knowing*. Chicago: Intellect: University of Chicago Press, 2011.

Benedek, Mathias and Emanual Jauk. "Creativity and Cognitive Control". In *The Cambridge Handbook of Creativity*. Kaufman, James and Robert Sternberg (eds.). Cambridge: Cambridge University Press, 2019: 200–223.

Bull, Cynthia Jean Cohen. "Sense, Meaning, and Perception in Three Dance Cultures". In *Meaning in Motion*. Desmond, Jane (ed.). Durham: Duke University Press, 1997: 269–287.

Carr, Jane. "Embodiment and Dance: Puzzles of Consciousness and Agency". In *Thinking Through Dance: The Philosophy of Dance Performance and Practices*. Bunker, Jenny, Anna Pakes, and Bonnie Rowell (eds.) Hampshire: Dance Books, Ltd., 2013: 63–81.

Dunagan, Colleen, Roxane Fenton, and Evan Dorn. "Modelling Improvisation as Emergence: A Critical Investigation of the Practice of Cognition". "Life Practices". In *Oxford Handbook of Improvisation in Dance*. Midgelow, Vida (ed). New York, 2019: 545–562.

Finke, Ronald, Ward, Thomas, and Steven Smith. *Creative Cognition: Theory, Research, and Applications*. Cambridge: MIT Press, 1996.

Gallagher, Shaun. *How the Body Shapes the Mind*. Oxford: Clarendon Press, 2006.

———. *Enactivist Interventions: Rethinking the Mind*. Cambridge: Oxford University Press, 2017.

Gallagher, Shaun, Lauren Reinerman-Jones, Bruce Janz, Patricia Bockelman, and Jörg Trempler. *A Neurophenomenology of Awe and Wonder: Towards a Non-Reductionist Cognitive Science*. New York: Palgrave Macmillan, 2015.

Gotlieb, Rebecca, Elizabeth Hyde, Mary Hellen Immordino-Yang, and Scott Barry Kaufman. "Imagination is the Seed of Creativity". In *The Cambridge Handbook of Creativity*. Kaufman, James and Robert Sternberg (eds.). Cambridge: Cambridge University Press, 2019: 709–731.

Grove, Robin. "Show Me What You Just Did". In *Thinking in Four Dimensions: Creativity and Cognition in Contemporary Dance*. Robin Grove, Catherin Stevens, and Shirley McKechnie (eds.). Melbourne: Melbourne University Press, 2005: 37–49.

Grove, Robin and Shirley McKechnie. "Introduction". In *Thinking in Four Dimensions: Creativity and Cognition in Contemporary Dance*. Robin Grove, Catherin Stevens, and Shirley McKechnie (eds.). Melbourne: Melbourne University Press, 2005: 1–8.

Hanna, Judith Lynne. *Dancing to Learn: The Brain's Cognition, Emotion, and Movement*. London: Rowman and Littlefield, 2015.

Iris Marion Young. "Asymmetrical Reciprocity: On Moral Respect, Wonder, and Enlarged Thought". In *Constellations* 3, no. 1 (1997): 340–363.

Ivcevic, Zorana and Jessica Hoffman. "Emotions and Creativity: From Process to Person and Product". In *The Cambridge Handbook of Creativity*. Kaufman, James and Robert Sternberg (eds.). Cambridge: Cambridge University Press, 2019: 273–295.

Leung, Angela K.-Y., Suntae Kim, Evan Polman, Lay See Ong, Lin Qiu, Jack Goncalo and Jeffrety Sanchez-Burks. "Embodied Metaphors and Creative "Acts"". *Psychological Science* 23, no. 5 (2012): 502–509.

McCarthy, Rosaleen, Alan Blackwell, Scott de Lahunta, Alan Wing, Kristen Hollands, Phillip Barnard, Ian Nimmo-Smith, and A.J. Marcel. "Bodies Meet Minds: Choreography and Cognition". *Leonardo*, 39, no. 5 (2006): 475–477.

McKechnie, Shirley and Catherine Stevens. "Visible Thought: Choreographic Cognition in Creating, Performing, and Watching Contemporary Dance". In *Contemporary Choreography: A Critical Reader*. Butterworth, Jo and Liesbeth Wildschut (eds.). New York: Routledge Press, 2009: 38–51.

McNeil, David. 1992. *Hand and mind: What gestures reveal about thought*. Chicago: University of Chicago Press.

Montero, Barbara Gail. *Thought in Action: Expertise and the Conscious Mind*. Oxford: Oxford University Press, 2016.

Plucker, Jonathan, Matthew Makel, and Meihua Qian. "Assessment of Creativity". In *The Cambridge Handbook of Creativity*. Kaufman, James and Robert Sternberg (eds.). Cambridge: Cambridge University Press, 2019: 44–68.

Rothfield, Philipa. "Chance Encounters, Nietzschean Philosophy, and the Question of Improvisation". In *Oxford Handbook of Improvisation in Dance*. Midgelow, Vida (ed). New York, 2019: 89–102.

Sternberg, Robert. "Enhancing People's Creativity". In *The Cambridge Handbook of Creativity*. Kaufman, James and Robert Sternberg (eds.). Cambridge: Cambridge University Press, 2019: 88–103.

Stevens, Catherine. "Minds in Motion: Dynamical Systems in Choreography, Creativity and Dance". *Tanzforschung*, 15 (2005b): 241–252.

Stinson, Susan. "Body of Knowledge". *Educational Theory* 45, no. 1 (1995): 43–55.

Tweney, Ryan. "Cognitive Science and the 'Dancing Brain'". In *Thinking in Four Dimensions: Creativity and Cognition in Contemporary Dance*. Robin Grove, Catherin Stevens, and Shirley McKechnie (eds.). Melbourne: Melbourne University Press, 2005: 149–153.

Young, Iris Marion. "Communication and the Other: Beyond Deliberative Democracy". In *Democracy and Difference: Contesting the Boundaries of the Political*. Edited by Seyla Benhabib, 120–135. Princeton University Press, 1996.

CHAPTER 3

Movement-Knowings: Affect and Embodied Intuition

Part of what makes embodied cognition theories distinctive is their emphasis on, or prioritization of, affect and, for some, its relation to kinesthesia. More specifically, according to embodied cognition accounts, affect itself is cognitive and is so, to a large extent, because of its relationship to the moving body. Embodied cognition theorists draw a distinction between many kinds of affect, such as vitality affect, primordial affect, and normative affect. These distinctions help one better understand the neurophenomenological and kinesthetic components of affect within the broader literature on embodied knowledge. In the phenomenology of dance, affect is central because of its role in embodied cognitive processes, such as kinesthetic empathy, embodied intuition, body memory, and blood memory. All of these embodied cognitive processes are only possible in, and from, movement. Movement gives rise to both the experience and the storing of these embodied forms of knowing; and it does so through the relationship between affect and movement.

In this chapter, I outline the theoretical trajectory of the research on embodied cognition and affect. If we understand the role of affect in consciousness as it relates to movement, we can make the connection between affect, movement, and embodied knowledge in phenomenological experience. For this reason, one can connect the phenomenology of affect to the

S. Welch, *Choreography as Embodied Critical Inquiry*,
https://doi.org/10.1007/978-3-030-93495-8_3

phenomenology of movement-knowings through the neurophenomenological approaches to affect and kinesthesia. As dancer Debhorah Hay rightly asserts, when dancing, the body itself uncovers its own wisdom and can then convey it to others (Sarco-Thomas 2019, 157). Accordingly, these accounts help relieve movers of the burden of the traditional presumption that embodied knowledge is enigmatic and mystifying when it (rightly) feels obviously not so. I examine and summarize why and how consciousness is tied directly to (the various modes of) affect and to clarify the role of affect in intuition for embodied knowledge and movement-knowings. If one can understand how affect is experienced in, and through, movement, they can have a better understanding of how to purposefully access and direct embodied knowledges. So when the body taps into the body as a form of embodied critical inquiry, the body can uncover one's specific normative relationships to the concept or theme under moving-consideration.

In "Affect: From Functionalism to Enactivism" section, I follow the role of affect from a more representational heavy version of embodied cognition to an account of embodied cognition that eschews representations—I trace from Damasio to Wetherell to Colombetti. My goal is to show the development of affect within embodied cognition accounts that advance from minimally functionalists accounts to more explicitly enactive models. In seeing the developments on this topic over time, one can see how commitments to affect in embodied cognition theory seem to become more and more radically embodied. In "Kinesthesia and Intuition" section, I outline the function of affect in the embodied forms of knowing that are highlighted in phenomenology and dance. I then articulate how it is that movement-knowings function with, and through—and are the dynamics of—creative movement practices, such as movement-as-curiosity. Because creativity and curiosity are the cornerstones of critical inquiry, it is helpful for me to show how movement-knowings can be made into specific practices of movement-as-curiosity through the directing of implicit but knowable embodied knowledge vis-à-vis affect.

Affect: From Functionalism to Enactivism

Antonio Damasio is one of the most well-known theorists who strongly incorporates the body into his account of cognition and affect; in most theories of embodied cognition, Damasio plays a visible role and many theorists position themselves in relation to him. While he has numerous works that adapt over time, one of his more popular pieces considers the connection between the mind and the self (2010). Like Lakoff and Johnson, his theory is seemingly reliant on representations in terms of his emphasis on the role of neural maps and mirror neurons. Like enactivism, his account is naturalist; but unlike enactivism, it is largely reductive to these biological processes, such as the neural maps. For these reasons, I would position Damasio within the functionalist camp of embodied cognition. Damasio regards the body as more of a vehicle for the brain's processes, rather than as the site. In this section, I want to start with a summary of Damasio's depiction of affect and show how two other theorists respond to his work, as they reconsider the relationship between the moving body and affect. I focus on his work in *Self Comes to Mind* since my work here is phenomenological in nature and thus starts from the autobiographical self that his theory of embodied consciousness terminates in.

Damasio's notion of self suggests that there is no distinction between the self-as-knower and the self-as-object. What is needed to establish subjectivity to give rise to consciousness is the incorporation of bodily functions, such as affect, into the role of the mind. For Damasio, subjectivity exists only because we have a particular kind of felt existence that animates or illuminates the representations we experience. As such, his notion of self, mind, and consciousness is grounded in a model of feelings-as-knowing. Consciousness, as he describes it, is merely a state of mind, one that is aware of, and has knowledge of, its own experiences in its surroundings. Embodiment is imperative because this knowledge of surroundings is a knowledge that is necessarily situated—and situatedness requires a body. This is because the knowledge in, and from, conscious situatedness is engendered through the various sensory modalities. He argues that our experienced sensations manifest as qualitative properties and that "[c]onscious states of mind are *felt*" (168).

Damasio argues that the body serves the mind's processes through its ability to exist in a sensory feedback loop with the mind. This feedback loop is comprised of the unending cycle between actions and neural maps/ movement and mind (68). The body experiences its surroundings and feeds this sensory information back to the brain where the brain then makes neural maps—and so representations—of these experiences to generate the knowledge of subjectivity needed for consciousness. These neural patterns map the sensory experiences of the external world as mental images and then remaps these images back out onto the world conceptually. These maps/images are the content that we pull from to engage in critical cognitive practices of manipulation and reasoning. Moreover, the neural maps inform the mind itself and are used materially to unconsciously guide motor behavior (67). Yet our neural maps are always adjusting and changing because we are always in constant motion. As such, it is not the mere body that gives rise to subjectivity—movement, specifically, is the foundation of the conscious mind (71). The body is central to his theory of consciousness and he explicitly backs away from a fully representationalist account of neural maps (68). However, Damasio's commitment to the nature of consciousness and subjectivity as a "state" of mind positions him on the weaker end of the embodied cognition spectrum.

To get to the autobiographical self, Damasio starts at the biological beginnings of the brain's protoself; the protoself is attached to the body via the body's primordial feelings. Primordial feelings are the images of the sensory experiences that are generated from the loop between the brain and the body. Primordial feelings are not limited to the sensory experiences that are mapped to and from the body. Damasio argues that even our non-felt images are felt because they almost always are accompanied by feelings (23). For Damasio, feelings are basic and constitute the amalgamation of our biological life regulative processes. Feelings that are the images of actions and emotions are the automatic program of actions (116); feelings are the perception that become executed in our neural maps (117). We have feelings that turn into emotions, that then turn into the actual feeling of the emotions, which is the (I would say, phenomenological, experienced) perception of what is happening in the body-brain feedback loop when we emote (ibid.). That feeling precedes emotion and drives the action plan as one becomes cognizant of the specific emotion; but the *feeling* of the emotion follows the emergence of the emotion. He explains that "[t]he actions are complemented by a cognitive program that includes certain ideas and modes of cognition, but the world of emotions is largely one of action carried out in our bodies..." (ibid.). One example he provides to illuminate the relationship between affect and cognition is that the feelings of sadness will slow down thinking—it will alter our action program—before we even realize we are sad, then this slowness of thinking makes us "feel" our sadness. His account of affect and emotions, though necessary to consciousness, is somewhat underwhelming but provides a good foundation for understanding how emotions and movement tie together for an understanding of the autobiographical self; and the autobiographical self is imperative to understanding an embodied critical inquiry since it is the autobiographical self on which we must reflect through movement.

Crucial to my purposes is his somatic marker hypothesis, which states that the especially valuable images of our situated experience are highlighted by emotion (186); the degree to which something triggers our emotions is an indicator for how important that experienced phenomenon is to us. And even more importantly, he argues that this felt knowing does not have to be explicit. On the somatic marker hypothesis, significance can feel like a gut feeling. While Damasio intends this hypothesis to help explain how our brain is able to sort through so many perceptions of images stored in the brain and then, further, to select certain kinds of images as necessary for survival, it provides us with an understanding of how somatics and emotion can pull experiences of significance to the surface of the autobiographical self, which then makes these experiences of significance available for us to reflect upon.

Margaret Wetherell (2012) provides an equally important account of affect, but one that largely rejects Damasio's trajectory of affect ➔ emotion ➔ feelings of emotion.[1] She explains that Damasio conceives of affect, and bursts of affect, as not especially interesting. For him, the subjective, phenomenological, psychological experience of emotion sits at the peak of biological and physiological body dynamics that operate to sustain homeostasis/harmony. Similar to my own concerns, Wetherell argues that Damasio's account of affect is too "primitive" since it appears to categorize affect as a basic urge or response that contributes to our body-brain-environment appraisal for action plans. One could say that Damasio undermines the potential power of his theory by merging a new theory of embodied cognition with an old school traditional theory of basic emotion theory. This results in a differentiation between affect and cognition that is not consistent with radical forms of embodied cognition (36–38).

Wetherell enters the embodied cognition debate about affect from a different direction—she situates her theory within a social context. Given

[1] She also rejects Brian Massumi, who is also well known for his work in the relation between the body and affect. However, his work is less concerned with the reductivist scientific perspective and more so in the social and political implications of affect. Massumi will be discussed at another time.

that embodied cognition theorists incorporate and emphasize culture and social interaction in their models of cognition, it seems fitting that she shift away from strictly physiological processes to gaze at affect from its social formation and its "psychosocial texture". To better understand affect as a cognitive experience, Wetherell looks at affective practices to show how affect is about sense and sensibility (14). She defines affect as both an embodied meaning-making and an emotion (4). She explains:

> Affect is always 'turned on' and 'simmering', moving along, since social action is continuously embodied. But affect also comes in and out of focus. The ongoing flow of affective activity can take shape as a particular kind of affective performance, episode or occasion, as in, for instance, a child's tantrum, a self-aggrandizing narrative, or a bounded experience of joy. Affective practices unfurl, become organized, and effloresce with particular rhythms. (12)

An analysis of affect such as this is significant for an account of embodied critical inquiry since this process requires that we use embodied affect to read bodied emotion and embodied metaphors.

Additionally, her conceptualization of social-affect practices aligns with Merleau-Ponty's notion of flesh. She explains that affective experience and existence helps us keep the world close to our skin (9).[2] The affective practices are coordinated and relational; they are, thus, distributed across this relational field, which one can equate with flesh, and each partner's contribution becomes meaningful—but only in relationship to the whole "affective dance" (87). The similarity between affect and flesh brings out her claims that affective practices are communicative and organized; she explains that they take shape and adjust as they "flow" in the discursive realm (15). Wetherell also has a notion of an affect loop but her loop concerns how affect becomes narrated through the spirals of the normative social flow to contribute to, and constitute, cognitive practices. She asserts:

> When we think about discourse in this practical light, distinctions between cognitive and non-cognitive, representational and non-representational, conscious and unconscious, language and the body become less and less clear-cut. Attention shifts to how affect is accomplished and ordered and moves away from adjudicating in abstract the exact relation between bodies and discourse, feelings and words. (53)

[2] See also: Suzanne Cataldi. "Affect and Sensibility". In *Merleau-Ponty: Key Concepts*. Diprose, Rosalyn and Jack Reynolds (eds.). Stockfield: Acumen Pub, 2008: 163–173.

Clearly, this understanding of affect is helpful to an embodied critical inquiry. Because our bodies are affective-cognitive-linguistic, and because our reflective practices are affective-cognitive-linguistic, an account of affect that highlights the role of the social in bodily physiological processes helps us draw a connection between inquiry and embodiment. It also brings out an important part of one's ability to use creative movement as a mode of communication and inquiry since it can point to the social-cognitive privilege affective skills and access can provide. Wetherell rightly explains that class is lived at the "gut level" via a "psychic economy" (111). Thus, it can shape a distinctive style of affect and affective relating that can privilege or disadvantage one in the exchange of flesh. Class is a large barrier to one's ability to cultivate, or even be exposed to, practices of creative movement that can open up access to embodied knowledge.

While she moves away from a representation heavy account of embodied cognition, she remains skeptical of non-representational accounts that fall on the radical enactivism end of the embodied cognition spectrum. She posits that we generate a process-representation if we look at them as modes of sense-making for the embodied cognitive processes of appraisal, perception, action, grasping, and evaluating (66). Wetherell, more specifically, redefines and relocates representation in a fashion suitable to her affective practice schema; representations under this frame are equally social-life discursive practices (76). By reformulating representations into discursive practices at the center of affective practices, she believes that we can retain a strong empirical tether between affect and embodied cognition instead of letting it float further into the theoretical domain. It may be that her account of affect would pair nicely with an account of action-oriented representations.

Giovanna Colombetti (2014) brings the two benefits of Damasio and Wetherell together in a naturalized phenomenological, enactivist approach. Regarding Damasio, she argues that his model of affect and emotion can merge with enactivism because his focus on the physiological balancing processes are inherently affective and evaluative in relation to the body-brain-environment admixture. But her concern with Damasio is similar to Wetherell's—he relies too heavily on basic emotion theory. But further, cognition cannot be fully reductive to neural maps; consequently, he does not allow for the possibility of significance for creatures that have no brain. While this question of whether or not brainless creatures have significance is rather far away from the question of whether or not dancing can be a mode of embodied critical inquiry, it does point to the deep problem that he separates cognition and emotion. For enactivism, this separation is impossible because its understanding of the mind is not dependent on the brain; for enactivists, the mind emerges from the creature's interactions with its world. While the enactivist centers the body's biological and physiological processes, the framework defines the enactive mind as a body—a body that perceives, lives, and acts; affect is not only integrated into processes of sense-making but is also constitutive of the mind. Because she is working from an enactivist approach, which is grounded in a dynamical systems framework, Colombetti swaps out the term emotion for emotional episode. Emotional episodes are dynamic patterns that are part of the self-organizing processes of the organism as a result of their interactions with their world. They are particularly dynamic because they entrain, or synchronize, all of the biological and physiological processes, and the mutual influences of the external environment, into tightly interwoven systems.

Colombetti works best for thinking about an embodied critical inquiry because she offers the most embodied account of affect. One might suggest that all one needs is access to the representations that follow from a functionalist operation of the body. But Colombetti's account can provide the full embodied account that equally includes both the normative and social factors that Wetherell rightly highlights. Colombetti regards affect as a sort of flesh ontology where affect can be passed on to others. Wetherell refers to this social experience as a contagion and Colombetti identifies it as a sympathetic sociability (172). However, Colombetti stops short of the conception of affect as practice. While I do not think that a theory of affective practices is incorrect, I do think that it is a complementary analysis of affect to one that accurately assesses the affect of the body. But most importantly, Colombetti asserts that embodied metaphors play a crucial role in revealing the dynamic and kinetic character of emotional experience (119).

Colombetti simultaneously posits that affective sensations are non-evaluative but are also ethical. In a sense, this aligns with Damasio in that her supposition matches his claim that the role of affect in situational appraisal is for an action plan. For her, the body can move from front and center to periphery or background depending on whether the subject's goal is reflective or pre-reflective; but always, the body is experienced as subject and never as object, even when one is assessing how their body is currently operating as a "medium for feeling-through" (117). She also differentiates between the bodily experiences of affect and emotion. Affect is kinesthetic and can be experienced as "action urges" insofar as it conditions perception in a way that binds the sensuous, sensing subject to the sensuous, sensible world (168). Emotion, on the other hand, is interoceptive and serves as the driving motivation behind behavior (166). For Colombetti, emotions are phenomenological because we live them but do not know them; moreover, emotions establish our relationship to the world (169)—emotions engender flesh.[3] But unlike other accounts, the body is most significant when we are engaged in activities strongly imbued with emotion, which indicates that the experience of emotion is radically embodied rather than radically psychological.

Kinesthesia and Intuition

From the above section, one can see how affect shares an important relationship with the ways in which movements engender and shape embodied ways of knowing. The trajectory from affect and embodied ways of knowing, when spelled out in more detail, moves from the role of kinesthesia in affect operation to the role of affect in generating movement-knowings, such as kinesthetic empathy, body memory, blood memory, and intuition—all of which can be accessed via somatic practices, which I discuss later. In this section, I demonstrate how these concepts merge and enfold into, and onto, one another in ways that produce embodied movement-knowings. I then discuss the specific and distinctive modes of operation of these forms of knowings in dance and how dancing can make use of them for embodied creativity practices, such as movement-as-curiosity.

Part of our kinesthetic interoception and proprioception is vitality affect. Vitality affects are differentiated, though not often strongly, from general affect. I should quickly warn that this distinction and relationship between the various kinds of affect is quite abstruse for the less-than-expert. However, these distinctions do play a role in how the body comes to generate and connect to its embodied knowledge; so it is important to include them and I do my best, as a less-than-expert, to properly articulate their relationship. Hanna explains that our interoception is how we

[3] Other accounts that really highlights the social and political account of affect are: Erin Manning. *Politics of Touch: Sense, Movement, Sovereignty*. Minneapolis: University of Minnesota Press, 2007; Brain Massumi. *Politics of Affect*. Cambridge: Polity Press, 2015. I take both of these accounts to work in tandem with the Merleau-Pontian conceptions of flesh and chiasm even more directly than Wetherell and Colombetti. I would like to point out that Massumi makes direct reference to Bergson. However, I do not think that they are fully and directly applicable to an account of embodied cognition and affect.

perceive our bodies, such that this contributes to how we experience emotions and our sense of well-being (Hanna 2015, 18). Colombetti says that vitality affects are most certainly a feeling, which I would associate with kinesthetic sensation, but primordial affect need not be an explicitly felt feeling (Colombetti 2014, 22). For her, and others, vitality affects are not "affects" in the way we usually conceive of affect as being directly tied to emotion, which as Hanna points out, is why interoception gives us insight to emotion but does not determine it. Colombetti rightly asserts that while affects come and go, vitality affects are always present. As such, they are deemed noncategorical feelings (ibid.). She actually associates vitality affects as being more associated with proprioception—following Stern, but she defines them in terms of kinesthetic experiences imbued with affective qualities, like muscle tension, muscle contraction, and fluidity that are brought on by changes in embodied physiological arousal via shifts in body dynamics (21); vitality affects are intrinsic to the bodied experience.

Excitingly, Colombetti remarks that vitality affects:

> ...come to the fore in bodily expressive forms of art, such as dance where bodily movements often convey specific qualities—of energy, explosiveness, lightness, heaviness, frenzy, tranquility, and so on. [(we can see the proprioceptive understanding of vitality affect here)] (21)

Additionally, and relatedly, she points out that vitality affects are also conveyed through touch and gesture. In making these remarks, she cites Sheets-Johnstone. Sheets-Johnstone equally draws a very tight connection between kinesthesia and affect (though not specifically vitality affect). She explains that the dynamic pairs of the phenomenological body are motion and emotion and kinetics and affect (Sheets-Johnstone 2009, 196). The link between interoception and proprioception is made evident here since she explains that our neuromuscular tension is always emotionally laden. For her, affect and kinesthesia can only be pulled apart during moments of reflection (202). To move from internal kinesthesia to external kinesthesia, Sheets-Johnstone posits the body's readiness to act—its felt urges—to be the origination of emotion. Concomitant with the previously discussed affect theorists, she argues that emotion operates as the affective motivation of our action plans—that is, emotion literally moves us from the inside out (ibid.). She claims that whether small or large, visible or not, felt or overlooked, emotions resolve into extensional or contractive movement (204).

When thinking about movement-knowings and the role of kinesthesia in interoception and proprioception, body memory becomes important because this is a form of embodied memory that strongly ties into how affect and vitality affect manifest. A body memory is not necessarily, or often, an explicit recollection.[4] Thomas Fuchs defines a body memory as the culmination of the embodied skills associated with patterns of movement and perception (Fuchs 2013, 10); he identifies six kinds of body memory: memory: procedural, situational, intercorporeal, pain, and traumatic.[5] But more commonly, body memory is understood to be the embodied experience of our lived past, which gives rise to particular dispositions to perceive and act in specific ways in certain contexts, especially in dispositions related to desire and communication (11). Fuchs' account is an enactive account; he argues that the body's experiences become anchored in the body memory, which gets spread out in ways that connect us to the environment "like an invisible network", and so gives rise to a kind of flesh that relates us to all things and people (ibid.). The forms of body memory he identifies is especially enactive. He states that body memory is not something that "represents" the past but is, instead, something that enacts the past; this form of embodied knowledge, like other forms of movement-knowings, becomes accessible through action. He explains that procedural body memory is constituted by our sensorimotor and kinesthetic faculties that are realized in dynamic processes, such as sequences of movement and skillful handling (12). He further explains that procedural body memory makes the world accessible and available to us. It is through this kind of body memory that the body can "reverberate" with what was ostensibly forgotten; consequently, the body carries the past around with it as a procedural field of possibilities (20). Situational body memory includes the inseparable units of body, sensory, and environmental perception; the senses work together to construct distinct impressions that become stored in the body (13). Intercorporeal memory includes an implicit relational bodily knowing of how to interact with others and elicit their attention (15). Pain-based body memories are inscribed directly into the body. Traumatic memory is one that is almost wholly

[4] See also: Michela Summa. "Body Memory and the Genesis of Meaning". In *Body Memory, Metaphor, and Movement.* Koch, Sabine, Thomas Fuchs, Michela Summa, and Cornelia Müler (eds.). Philadelphia: John Benjamins Pub. Co., 2013: 23–41.

[5] It is worth noting that in his articulation of body memory, Fuchs draws from both Merleau-Ponty and Bergson.

buried in the body memory and sequestered from the larger sense of meaning one has of their self.

Sheets-Johnstone offers another useful kind of body memory that is strictly kinesthetic—kinesthetic melody. Kinesthetic melodies contain the affective dimension of movement and then informs that movement; kinesthetic melodies contain the signature of our qualitatively felt movement patters that generate "synergies" of meaningful movement (Sheets-Johnstone 2012, 43). Because kinesthetic melodies are affective dynamic body memories, an impulse of a movement-based kinesthetic memory can activate the full sequence of movement remembered alongside it. We would call such movement sequences particular activity skill sets—kinesthetic melodies might be described as a sort of robust affective sensorimotor contingency that grounds some kinds of movement-knowings. She argues that meaning itself resides in, and emerges from, the kinetic dynamics of dance. She rightly avows:

> While dances may be narrative—while they might be dramatic in nature or tell a story—the drama or story is kinetically engendered in the sheer qualitative dynamics of movement. There may be connotative gestures, even denotative ones, but the dance's qualitative kinetic dynamics carry the day. They are the "stuff" of the dance, its heart, and, for the dancer, it's very breath. The dancer is not moving through a form; the form is moving through her. (52)

The moving body generates meaning through creative, dynamic movement in a way that produces and communicates a movement-knowing that is characteristic of enactivist understandings of cognition. Although Sheets-Johnstone remarks that kinesthetic memories and melodies are non-cognitive, she insists that the kinesthetic is inherently tied to affect and they have dynamic vitality affect dimensions. I would surmise that her designation of kinesthetic consciousness-modes of experience as non-cognitive are made in the absence of a nonrepresentationalist framework from which to work.

I hold body memories and kinesthetic melodies are expressions of general kinesthetic intuition and perception. Kinesthetic intuition is central to an embodied critical inquiry because it allows us to utilize our bodies and kinesthetic movement as a mode of perceiving affordances that our other sensory modalities have been trained to ignore. As previously discussed, Sheets-Johnstone follows Langer quite strongly on the non-cognitive understanding of the kinesthetic. Because she (rightly) believes that movement is taken up directly and immediately, she moves towards Langer by asserting that immediate uptake occurs without "reason, language, or thought" (54). This immediacy of comprehension is consequent of the kind of kinesthetic perception and intuition that Langer posits as crucial to the comprehension of art generally.[6] She argues:

> Artists and art-lovers generally agree that artistic perception is intuitive. It takes place, they would say, spontaneously and immediately, without reasoning, without the benefit of logic. Some would claim, furthermore, that a special power of intuition leads the art connoisseur to knowledge of an inner reality, which the philistine can never know. This knowledge, they say, comes through feeling, not thought; it is irrational; it is a metaphysical contact with the real…The great difference between rational insight and artistic insight lies in the ways intuition is elicited. (Langer 1957, 60, 69)

Langer is right to insist that artistic perception is intuitive and comes from direct insight, which also means that it is not a function of discursive thinking. But she is incorrect in this claim from an enactivist perspective, since her conception of discursive thinking excludes the kinds of body logics that exist across various embodied cognition accounts. She presumes that embodied and kinesthetic intuition cannot involve belief since it does not involve proposition. This misses what the relationship between affect and body memory tells us—the body itself is the vessel of many beliefs and norms. And, as I am arguing, we can access them through the kinesthetic perception and intuition of creative modes of movement, such as dance. Of course, this requires removing the separation Langer makes between the mover and the perceiver.

Susan Melrose (2009) offers an interesting and informative account on the particular "expert" intuition of dancers, which she conceives as a sort of sensibility indicative of an expert memory. For her, expert/dancer intuition is likely an advanced embodied knowledge of movement that processes information during the uptake of movement that occurs more swiftly than a non-expert dancer. What she seems to suggest is that this dancer intuition is an "intuition"; that is, she seems to suggest that what dancers regard as a movement/moving intuition is really just highly advanced executive know-how. But this understanding of embodied intuition isn't directly at odds with the general embodied cognition account of

[6] For more on kinesthetic intuition, see: Michela Summa. "Enacting Perception: The Relevance of Phenomenology". *Paradigmi: Rivista di Critica Filosofica* no. 2 (2014): 97–115.

sensorimotor contingencies and affective sense-making capacities as being the source of embodied intuition. Embodied intuition is the ability of our bodies to hark back to a bodily implicit remembered know-how that can lead us in decision making, action plans, affective assessment, and creative ventures, especially with respect to constructing choreography. Following G. Ulmer, Melrose tells us:

> What suddenly 'fits' seems to have been conjured up out of a nowhere of knowledge and a nowhere of individual intent: she cannot make it appear, but can bring to bear, with a semblance of patience, the various schemata that apply to the professional creative task, and wait. It—the intuitive 'product' (Ulmer 1994)—is likely to be felt by her, in its full impact, as something fragile, new, but 'right'.... (33)

She cites Ulmer to explain that dancer intuition is a Gestalt mold that crosses sensory modalities in a way that cannot be abstracted from the affective feeling body (34). This specific understanding of expert intuition in dance is precisely the kind of movement-knowing seems to be an amalgamation of body memory and kinesthetic melodies that makes an embodied critical inquiry possible.

In reference to improvisational movement, Hilda Wengrower (2016) points to the efficacy of creative movement to tap into, and draw out, embodied knowledges—particularly body memory—within our cognitive unconscious by allowing the mover to abandon expected movement directions to discover new "psychophysical forms" (16). To understand how this is possible, it is helpful to look at the work of Joan Chodorow (2016) on dance movement therapy. We are able to uncover new psychophysical forms tied to movement-knowings because our direct experience is multisensory and so forges a strong relationship between emotions and our imagination. She explains that our interoception is integral to our emotional processing in the cultivation of body awareness and the self. She proffers the notion of the kinesphere to help explain the holistic nature of our multisensory image-experiences that contain infinite possibilities for embodied action and interaction (59). However, it is important to recognize the intermediary role for embodied intuition between interoception processing of bodied-self-awareness and the recognition of affordances for embodied action. For as Albright avows, a responsive body is one who can fully trust in its own intelligence of reflex (Albright 2019, 27).

Langer's discussion of artistic perception as a mode of intuition is helpful for thinking about action and meaning-making affordances. There are two ways this is possible. The first is the most obvious; if things like affective kinesthetic melodies are a part of embodied intuition, then embodied intuition can be a significant driving factor—concomitant with our kinetic body logics—in uncovering and exploring physical movement affordances. As Albright rightly claims, when we shift our focus from the visual to the sensual, which requires that we simultaneously open ourselves to the world while grounding our self-awareness, we can learn to move with a kinaesthetic curiosity (Albright 2019, 26), which I would identify as an intuitive embodied knowing. In short, artistic perception and intuition help clarify the content of movement-knowings when movement is specifically used for movement-as-curiosity. They do so precisely through the identification and exploration of the physical affordances that the body affectively selects as valuable, and pursues, for the purpose of sense-making. Second, Langer's depiction of artistic perception and intuition reveals how movement can access embodied knowledge because these kinds of kinesthetic capacities aid us in the reading of the movement itself. What I mean by "reading movement", which I will discuss later, is the activity of using all sense modalities to reflect on previous movement to grasp and attempt to comprehend what, and which, knowledge(s) and meaning(s) that specific movement contains (not just displays). Embodied artistic perception and intuition undergird embodied intuition from the inside out and from the outside in. Our embodied intuition guides us where we need to go, and it also tells us what we need to know as a result of where we needed to go. Procedural memory, which is our kinesthetic and sensorimotor faculties that make the dynamic world dynamically available to us, is operative in the relationship between movement-knowings and embodied intuition in both directions; procedural memory is what illuminates the affective and physical affordance possibilities available to us. The ability of embodied intuition to permeate the literal flesh and carry knowledge between the outside (affordances) and the inside (affect) is why it is the core of a process of an embodied critical inquiry.

Kinesthetic empathy is also advantageous to movement-knowings. While kinesthetic empathy may not seem to have a particular role here since the method of embodied critical inquiry does not involve an audience (outside of a facilitator/witness), it does play a significant affective role in partnering activities. A sensing body, and its dynamics, such as tension, fluidity, and force, helps the moving body be read by another because kinesthesia is—as Langer points out—part of perception and intuition. Reason and Reynolds (2010) highlight that there is a strong link between emotion and movement because our sense modalities report movement that is connected to parts of the nervous system that is related to affect (53). Therefore, kinesthetic responses to movement, often times triggered through mirror neurons, will trigger the corresponding affect: tense triggers tense (54; Katan-Schmidt 2016; Welch 2019). When thinking of audiences, Reason and Reynolds explain that the emotional response to movement is produced by the process of engaging the work as a whole through interpretive strategies (68). But kinesthetic empathy also works between movement partners since this can help them merge their embodied intuitions. That is, as I will show later, through certain kinds of contact practices, partners can connect kinesthetically-affectively in ways that merge embodied intuition and allows them to have shared embodied communications. The dynamics of movement can bleed into and between embodied partners precisely because various aspects of their nervous systems are linking through the overlapping of their body-environments, which can make for a body-brain-environment that encompasses both. This is turn, through the exchange of affect and perception via mirror neurons and vitality affect, can create a bonded pair that functions simultaneously as reciprocal witness and internal dialogue embodied interlocuter.

Movement-knowings are the foundation of an embodied critical inquiry process and method. Processes and practices of inquiry are purposefully reflective activities; and what often goes unnoticed is the depth to which cognition is embodied, such that we become able to engage in critical personal reflection directly through our bodies. The operations of sensorimotor contingencies in sense-making, norm construction, and norm assessment, along with the deposits of bodied knowledges and memories, create a route through which we, as body-minds, can engage and reflect upon the movement-knowings inside us through movement itself. That some might be skeptical about the ability of individuals to access kinds of knowings that belong to the body through the body seems ironic, at best. Certainly, if moving bodied experiences can insert knowledge and valuation into the flesh then moving bodied experiences can pull it back out. This is one of the important roles that enactivism, as a more radical form of embodied cognition, does for this account. Instead of having to suppose that actualities such as body memories somehow turn into mental representations that can only be understood through extracting them and articulating them into propositional context seems to be entirely at odds with any framework that acknowledges and values embodied affective knowledges. What these enactive frameworks provide is a means through which to understand the actuality of such embodied affective knowledges and the methods and practices through which we can access and engage them reflectively.

Bibliography

Albright, Ann Cooper. "Life Practices". In *Oxford Handbook of Improvisation in Dance.* Midgelow, Vida (ed). New York: Oxford University Press, 2019: 25–36.

Cataldi, Suzanne. "Affect and Sensibility". In *Merleau-Ponty: Key Concepts.* Diprose, Rosalyn and Jack Reynolds (eds.). Stockfield: Acumen Pub, 2008: 163–173.

Chodorow, Joan. "Dance Therapy: Motion and Emotion". In *The Art and Science of Dance/Movement Therapy: Life is Dance.* Chaiklin, Sharon and Hilda Wengrower (eds.). York: Routledge Press, 2016: 53–74.

Colombetti, Giovanna. *The Feeling Body: Affective Science Meets the Enactive Mind.* Cambridge: MIT Press, 2014.

Damasio, Antonio. *Self Comes to Mind: Constructing the Conscious Brain.* New York: Vintage Books, 2010.

Fuchs, Thomas. "The Phenomenology of Body Memory". In *Body Memory, Metaphor, and Movement.* Koch, Sabine, Thomas Fuchs, Michela Summa, and Cornelia Müler (eds.). Philadelphia: John Benjamins Pub. Co., 2013: 9–22.

Hanna, Judith Lynne. *Dancing to Learn: The Brain's Cognition, Emotion, and Movement.* London: Rowman and Littlefield, 2015.

Katan-Schmidt, Einav. *Embodied Philosophy in Dance: Gaga and Ohad Naharin's Movement Research.* London: Palgrave Macmillan, 2016.

Langer, Susanne. *Problems of Art: The Philosophical Lectures.* New York: Charles Scribner's Sons, 1957.

Manning, Erin. *Politics of Touch: Sense, Movement, Sovereignty.* Minneapolis: University of Minnesota Press, 2007.

Massumi, Brian. *Politics of Affect.* Cambridge: Polity Press, 2015.

Melrose, Susan. "Expert-Intuitive Processing and the Logics of Production: Struggles in (the Wording of) Creative Decision-Making in 'Dance'". In *Contemporary Choreography: A Critical Reader.* Butterworth, Jo and Liesbeth Wildschut (eds.). New York: Routledge Press, 2009: 23–37.

Reason, Matthew, and Dee Reynolds. "Kinaesthesia, Empathy, and Related Pleasures: An Inquiry Into Audience Experiences of Watching Dance". *Dance Research Journal* 42, no. 2 (2010): 49–75.

Sarco-Thomas, Malaika. "Mass May be the Single Most Important Sensation: Perceptual Philosophies in Dance Improvisation". In *Oxford Handbook of*

Improvisation in Dance. Midgelow, Vida (ed.). New York: Oxford University Press, 2019: 151–169.

Sheets-Johnstone, Maxine. *The Corporeal Turn: An Interdisciplinary Reader*. Charlottesville: Imprint Academic, 2009.

———. "From Movement to Dance". *Phenomenology of Cognitive Science* 11 (2012): 39–57.

Summa, Michela. "Body Memory and the Genesis of Meaning". In *Body Memory, Metaphor, and Movement*. Koch, Sabine, Thomas Fuchs, Michela Summa, and Cornelia Müler (eds.). Philadelphia: John Benjamins Pub. Co., 2013: 23–41.

———. "Enacting Perception: The Relevance of Phenomenology". *Paradigmi: Rivista di Critica Filosofica* 2 (2014): 97–115.

Welch, Shay. *The Phenomenology of a Performative Knowledge System: Dancing with Native American Epistemology*. New York: Palgrave Macmillan 2019.

Wengrower, Hilda. "The Creative-Artistic Process in Dance Movement Therapy". In *The Art and Science of Dance/Movement Therapy: Life is Dance*. Chaiklin, Sharon and Hilda Wengrower (eds.). New York: Routledge Press, 2016: 13–32.

Wetherell, Margaret. *Affect and Emotion: A New Social Science Understanding*. London: Sage Pub., 2012.

CHAPTER 4

Somatics: Improvisation and Feeling-Practices

While I have demarcated the domain of interest within dance to that of the phenomenology of dance, I have been using the term "dance" a little fast and loose. It is not that I have been using the terms with some very specific form or mode of dance in mind, but I have not been making clean and pertinent distinctions between two important processes within dance creation: improvisation and choreographing. I also use these two terms loosely. I designate exploration and experimental practices of creative movement as improvisation; I designate the purposeful building and sequencing of movements and transitions as choreographing practices. Both improvisation and choreographing play decisive roles in the embodied critical inquiry process, when taking the processes as a whole from beginning to end. However, the stage of inquiry, as opposed to the whole process from inquiry to narrative, rests on improvisational movement practices. It is through improvisational movement that movers are able to access embodied knowledges that I mark as movement-knowings.

I identity the process of embodied critical inquiry as choreography, rather than improvisation, for two reasons. First, the practice of innovative, creative choreography construction regularly involves some engagement in practices of improvisation—again, loosely construed; thus, improvisation can be said to fall under the umbrella of choreography as a practice even though it is a distinct stage within the choreographing process. Second, because I posit purposeful inquiry as being constitutive of a specifically critical inquiry, there is a kind of expectation that the goal of

S. Welch, *Choreography as Embodied Critical Inquiry*,
https://doi.org/10.1007/978-3-030-93495-8_4

inquiry is to have a response to the questions raised—or at least some form of consideration of the questions uncovered. In this process of embodied critical inquiry, the process of constructing a "declarative", narrative movement "piece" is an end goal (but it is not necessary). That is, the mover goes through the improvisational processes to access embodied questions they seek to draw out and then construct a response to those questions with, and through, choreographing.

The purpose of this chapter is to show how somatic and improvisational practices are substantial embodied movement practices to mine the body for questions.[1] To do so, I highlight some key practical components of both improvisation and choreography that anchor some of the theoretical claims in the previous chapter. These clarifications then carry through to aid in the theoretical analysis and explication of the embodied critical inquiry itself in the next chapter. In the first section of this chapter, I set out some core foundations for both improvisation and choreography and some key pertinent distinctions between them. These differences are important because they help illuminate how and why each of them plays discrete but equally important roles in the choreography process. In the second section, I focus on the most important aspect of dance to the embodied critical inquiry process: somatics. I first give a general overview of what somatics is and what sorts of practices constitute, and contribute to, successful somatic practices. I highlight the use of somatics in dance movement therapy to better illustrate how somatic practices function with respect to embodied-affective knowledges. In the final section, I translate the movement-knowings in the previous chapter into these somatic practices; or rather, I show how these somatic practices are excellent practices for the movement exploration needed to purposefully access embodied knowledges for the querying process from an enactivist perspective.

Choreography and Improvisation

The ability of choreography to be a mode of research is becoming established. Areas such as performance philosophy, dance research, artist research, and performance research all emphasize the capacity of

[1] For a thorough and historical description of the various somatic instructors, techniques, and uses, see: Martha Eddy. *Mindful Movement: The Evolution of the Somatic Arts and Conscious Action*. Chicago: Intellect, 2017.

movement and performance, generally, and dance, specifically, to be methods through which to conduct research and, equally, to disseminate its findings. It is becoming more commonly recognized, then, that dance is a method through which one can literally philosophize. While some philosophers have used dance as a metaphor for doing philosophy, others have pointed to its real potential to be philosophy itself. Along with the dance phenomenologists I've been engaging, Nietzsche is another well known (though debated) figure who believed in the philosophical power of dance; Senghor, though less recognized and appreciated, was a proponent of philosophical dance. However, until very recently, such scholars' musings about dance have typically been regarded by the mainstream of academic philosophy as fanciful, metaphorical, or highly subjectivist. Now, with new scholars all over the world producing exceptional dance research theory and pedagogy, the push for dance to be acknowledged as an actual and direct method of doing philosophy with the bodymind is receiving less skepticism.

Efrosini Protopapa (2013) offers an exciting articulation of how choreography is now being recognized as a means through which to conceptualize itself—that is, how to use choreography to produce concepts about, and within, dance. She explains that there is a sort of conceptual cycle involved in practices of choreography. Dancers go into the choreographing process with concepts in mind and then, through their movement practices, are able to produces concepts specifically about dance. In other words, going into choreographing with concepts allows the dancer to come out on the other end with equally informative concepts to understand choreography itself. Protopapa states:

> ...I would argue, not only does dance exercise the philosophical, but it is allowed to start discovering and generating original modes of thought from within its own complex economy of practice (279) ... If this last proposition of works operating as philosophy has something to do with choreographers' self-reflexivity inventing concepts of performance through the very practice of performance-making, then this idea indeed corresponds perfectly to Deleuze and Guattari's description of philosophy as 'the art of forming, inventing, and fabricating concepts' (Deleuze and Guattari, 1994, pg. 2). (284)

On her account, philosophical theorizing through dance is an act of thought where concepts are created as one thinks in movement and so are presented as a becoming (ibid.). Dance and other forms of performance are performances of philosophical thinking precisely because the thought and conceptualizing are actively happening in us as we move (288). She posits that performance is a thinking because the mover is performing thinking. It is the process and performance of choreography practices that contain distinctive philosophical modes of thinking and conceptualizing that occur within, and through, the body directly; this form of philosophizing is immediate and does not require that the performer take their embodied theorizing from the floor to the desk.

Improvisation is a stage in this choreographing process that I believe is a mode of philosophical thinking that does not involve a *performed* thinking.[2] On my account, improvisation functions specifically as the philosophical query rather than the philosophical construction that yields conceptual products. I think that Camille Buttingsrud (2021) and Vida Midgelow (2015) provide the most accurate and detailed depictions of improvisation that carves out what I propose to be dance's querying potential. I purport that these authors are both working squarely within the realm of embodied critical inquiry from a dance perspective. Their work is pivotal in recognizing the powers of dance to be philosophical in nature and philosophical in practice. While some might suggest that performance philosophy is merely a tag to mark performance theory works as philosophy-adjacent, works such as these stand out as dance paradigms of explicit performance philosophy.

Buttingsrug posits that, through improvisation, dancers cultivate embodied knowledges, and they are able to know when embodied knowledges are accessed and working in the movement practices. Improvisational practices allow dancers to reflect non-conceptually through their affective activities and this non-representationalist reflection is what unveils their embodied knowledges to them. On her account, improvisation, as a distinct mode of embodied sense-making is a discrete state of consciousness that is not derivative, reductive, or dependent on any other state of consciousness; she terms this improvisational state of consciousness as a state

[2] See also: Susan Foster. "Taken By Surprise: Improvisation in Dance and Mind". In *Taken By Surprise: A Dance Improvisation Reader*. Albright, Ann Cooper and David Gere (eds.). Middletown: Weslyan University Press, 2003: 3–7; Joel Gluck. *Insight Improvisation: Melding Meditation, Theater, and Therapy for Self Exploration, Healing, and Empowerment.* (Self-published) 2017.

of embodied reflection (Buttingsrud 2021, 3). With respect to improvisational dance, she explains that:

> What the dancers describe, though, is a bodily directedness that is both deliberate and immediate, which seemingly effortlessly, encompasses various facets of the situation simultaneously…The attention of this bodily reflective state embraces the whole context of the activity. This form of bodily intentionality is not only multifaceted, but seems altogether different from what we know from established philosophical theories. It does not have the directedness conceptual reflection has towards its object of reflection, where the attention moves like a pointing finger towards the object, and narrows the field of focus down to one specific end. Neither does this bodily intentionality work on a pre-reflective level as our motoricity does—it is far too attentive, fine-tuned, and deliberate to fit that description. (7)

Similarly, Midgelow, in evidencing performance as a mode of research, directly takes up the inquiry of how, exactly, dancers come to know through the body. She explains that there is a specific body ontology that contains all of our embodied knowledges and concepts that can surface during improvisational movement practices in reflexive and critical ways (Midgelow 2015, 109). The purpose of her paper is two-fold. First, she inquires as to how improv can be an embodied mode of research and way of knowing. To answer this question, she further asks how a dancer can know something of the dancing body in action; that is, she asks how we can come to know through the body using specifically improvisational movement (111). Second, she questions how it is possible for viewers to make sense of this embodied knowledge that can be displayed on the flesh of the dancer's body in performance.[3] With respect to the first question, she avows that we are able to know through the body because we can activate our lived kinaesthetic consciousness through improvisational movement. What Midgelow does that is most significant is explain our ability to directly access the knowledge in the body through the use of the body in terms of inner logic, which I believe would be similar to Sheets-Johnstone's kinetic body logic. According to Midgelow, our thematizing of our inner logic is necessary to activate Practice as Research:

[3] One can imagine Langer's notion of virtual power as being relevant to discussions regarding how the meaning of improvisation can be taken up by an audience. See: Colleen Dunagan. "Dance, Knowledge, and Power". *Topoi* 24 (2005): 29–41.

> Through listening to these experiences (from) within the act of dancing, clustered themes begin to take shape forming an inner logic. Taking account of experience (which incorporates experiencing physical, sensate, emotional, philosophical and conceptual) allows the phenomena of improvising to be articulated and understood 'as actually had rather than laying some invented theoretical scheme on experience' (Gendlin in Todres 2007: 27). (119)

Her motivations and arguments are strikingly similar to my own. Though other performance philosophers point to our ability to tap into embodied knowledge through dance—specifically improvisation—Midgelow has taken the power of improvisation further to discuss its capacity to purposely draw this knowledge out of the body. While Buttingsrud is correct in her explication of improv as a non-representationalist practice, Midgelow shows how improv can switch from a pre-reflective practice to a reflective practice that can translate the embodied movement-knowings into reflective connotative knowledge. It is precisely through reflection on improvisational movement, which—in my embodied inquiry process—happens at the video reviewing phase, that embodied concepts can be plucked out of the raw improvisational experience into articulate forms. She tells us, correctly, that improv is a self-revealing form. In her work, as in my work here, she argues that prioritizing the work in improvisational practices can help us utilize choreographic strategies to make the "interior become visible" (129).

As I have stated, improvisation and choreography—though tightly connected—can come apart when actively and purposefully aiming to utilize the discrete expressive capacities of each stage of dance creation. Although it is important to heed Albright's warning against parsing them according to process and product (Albright 2019, 26), it is beneficial to recognize the possible differentiations between them. As can be gleaned from above, one distinction is that between the pre-reflective and reflective consciousness relationships to the movements. Improvisation emerges from a kinaesthetic engagement with embodied knowledges and these practices cull embodied knowledges through intuitive movement. Once this movement emerges, the dancer can reflect on the movements. Lynne Anne Bloom and L. Tarin Chaplin (1982) tell us that improvisation serves as the preparation stage from which choreography can grow since these forms of practice are ways of "putting analyzed, theoretical material into the soma-psyche (body-spirit)" (5). They explain:

> Improvisation is a way of tapping the stream of the subconscious without intellectual censorship, allowing spontaneous and simultaneous exploring, creating, and performing. Improvisation emerges as an inner-directed movement response to an image, an idea, or a sensory stimulus. (6)

This leads to the second way in which improv and choreographing can be separated—intention. They assert that choreography is brought into the world via our senses and gut (ibid.). So then one might say that the intention behind improvisational practices is to bring out movement-knowings using embodied intuition for the purpose of developing choreography. The intention behind the construction of choreography, through the manipulation and synthesis of the improvisational movements, is to communicate with others. In short, the intention behind improv is to come to know and the intention behind choreographing is to tell what one comes to know.

Bloom and Chaplin hint at this division of intention when they speak to the blurriness of intention during the holistic process of choreographing. They say:

> Sometimes there isn't a clear intention; there is only an inner drive, a restless energy, vague and undirected, a need to create. And sometimes for a while that vague nudge has no clearly defined, identifiable intent or theme. Often the choreographic process itself includes discovering and defining the theme and intention. The choreographer goes to the studio and begins to work, and in the working, in the moving, something happens, something connects, something becomes important, and almost on their own, the theme and intention self-clarify (9) …It is clarity of intention at an intuitive or rational level which helps you to determine which material fits and which can be thrown away. (10)

Thus, improvisation seems to be an intentional mining of the body using embodied intuition to extract movement-knowings; and the construction and piecing together of movement, shape, and dynamics seems to be an intentional act to execute those findings. Larry Lavender (2009) refers to such improvisational intentions as "dance-making intentions" (74) and "aboutness" intentions (75).[4] Just as I suggest that the dancer

[4] For an account of how the body itself, rather than the dancer, generates a kind of intentionality, see Rothfield 2019; here she conceptualizes the forces internal to improv through the will to power.

may move from improvisational intuitive movements to articulated choreographic movements by reviewing them, Lavender similarly suggests that this understanding of improvisational movement can be achieved with the assistance of a facilitator. As I will show in the "results" chapter of the book, my dancers combined individual video witnessing with facilitation from their movement witness to cultivate their explicit knowledge of their intuitive movements.

This division of intentionality between improvisation and choreography gives rise to the third distinction between them. We can say that inquiry is a subjective process and (narrative) choreographic construction aims to be objective.[5] This is just a translation of my claim that improv is a process of becoming and coming to know whereas choreographing is outwardly communicative. But seeing the distinction articulated in this way highlights, again, the explicit aims for each stage of the process. Certainly not all dancers are going about the choreography process this way and not all dancers are aiming to produce narrative pieces; but when they do, this is one way we can understand what is going on at various points in the method. This distinction between the subjective phase and the objective phase is central to the method of an embodied critical inquiry since the mover will have to communicate with herself internally-intuitively and then make sense of what she comes to know so that she can purposefully communicate her knowledges in ways that make sense to others.

Somatic Work

In this section, I lay out the role of somatic work, since I take this to be the most effective priming activity for engaging embodied intuition via kinesthetic activation. The term somatic comes from the term soma, which refers to the living body in its wholeness; or, as it is described within neurobiology—the body of the precognitive self (Fraleigh 2015a, xxi) since it is the site of both perception and self-perception. Our "soma" is responsive to all internal and external stimuli and our ability to perfect somatic practices allows us to account for all of the ways in which our body and its functioning contributes to our overall well-being in our lived experience. Somatic work is comprised of various techniques and movement practices that take account of, and employ, the holistic body, which is constituted

[5] For a detailed analysis of how narrative choreography can be made objective, see: Welch 2019.

by the body, mind, spirit, and its environment (Brodie and Lobel 2012, 6). Through somatic work, which consists either of stillness-movements or improvisational movements, one can learn how to recognize and exploit the connections between our interoception, proprioception, kinesthesia, and exteroception, which constitute the internal network of pathways of, and for, bodied communication.

Martha Eddy explains that somatic work helps convert the subconscious into something that can be shared and, thereby, engenders a link to conscious communication (Eddy 2017, 13). One of the most beneficial additions to somatic work that can help movers establish this link is the presence of a witness. The witness can serve in an emotional support role or as a facilitator to the movement exercises. The witness can represent the ever-present condition of intersubjectivity and social embeddedness even when we turn our awareness and attention inwards. It is possible, though, for the mover to serve as her own witness using self-reflexive practices and dialogues.

Somatic work is utilized in divergent movement domains. For example, somatic work can be employed fully on its own so that one can learn body awareness through techniques of "quiet consciousness" and connect to the wisdom of the body. One of the most recognized forms of somatic movement practices developed in dance to achieve this goal is termed Authentic Movement.[6] For this reason, many somatic techniques are developed in relation to the notion of felt sense. Amanda Williamson (2018) draws from Gendlin in her account of somatic practices that include: silent-sensing, tending to and listening for a felt sense, tending to and listening for a felt-sense, and moving from a felt sense (84). Here, somatic work might be regarded as parallel to meditative practices, except somatic practices are directly concerned with the quiet kinesthesia (active non-action) of the body. Somatic work is also used in physical therapy contexts. Patients can acquire and practice somatic practices that get them more aligned with their body awareness so that they can better tune into their bodies to see how and where certain injuries need special attention. In this context, the use of somatic work is directed towards achieving improved body functionality and prevent further injury. These goals are achievable because somatic practices allow one to access various manners of embodied knowledge (8).

[6] These movement practices are developed and advocated by Joan Chodorow, Mary Stark Whitehouse, and Jane Adler.

Fraleigh (2015b, 2019) is a leading dance philosopher on the relationship between improvisation and somatic work; she explicates the relationship between affect and somatics to help others understand how, and why, movement can be deployed to enhance perception and emotional responsiveness. As Williamson pulls from Gendlin, Fraleigh pulls from Damasio's conception of the somatic marker to point to our implicit consciousness, which constitutes the soma. She purports that we access our implicit body knowledge using a mode of somatic perception, which is the form of perception oriented by sensate content. Since our perception and emotional experiences are tied to affect, Fraleigh argues that the cultivation of somatic techniques can give us more options and choices on how to respond affectively to situations and others (5) because somatic movement allows our subconscious to "speak:

> Descriptions of immediate experience are a way of sharing somatic experiences, but finding the direct and intuitive way to describe movement, affect, and our sensate proximity to others is at first daunting. It requires one to voice what is not initially discursive, but kinesthetic in nature, felt as kinesthetic dynamic.... (21)

As she says, motion is part of e-motion (7). These somatic techniques provide our movements with affective and aesthetic purpose; on her account, making movements from a distinctively joyous somatic place produces a form of kinaesthetic art (ibid.). Similar to my claims regarding the distinction between improvisation and choreography in the above section, Fraleigh highlights the split between implicit somatic movement and narrative work. Somatic movement appears to be pre-reflective insofar as she posits that practices of reflection can affirm what we learn somatically. As such, she claims that narrative can flow more effectively following somatic movement practices. It is in this way, she espouses, that Practice as Research greatly benefits from somatic work (19).

Given the connection between somatics and affect, one of the most recognized connections between our soma and the body's implicit knowledge is the body memory.[7] This is why somatic work is the cornerstone of affect impelled improvisation and dance movement therapy practices.

[7] For an account on the relationship between culture and collective body memory, see: Thomas Fuchs. "Collective Body Memories". In *Embodiment, Enaction, and Culture: Investigating the Constitution of the Shared World*. Durt, Christopher, Thomas Fuchs, and Christian Tewes (eds.). Cambridge: MIT Press, 2017: 333–352.

Sharon Chaiklin (2016) explains that DMT is based on the presumption that individuals can express their own impulses, in relation to themselves and in relation to others, through dance (5). One prominent goal of dance movement therapy is to make use of somatic work and creative, intuitive movement to help individuals access the memories that highly emotional traumatic experiences leave in, and on, the body so that they may assess and reformulate their relationships to these experiences. However, trauma and PTSD, while well-known contexts for the use of somatic movement, are not the only ones. DMT can also aid individuals in positive character trait development and expression, as well as the formation of joyous practices.

Because kinesthesia and affect cannot be separated, dance movement therapy serves as a primary practice for gaining insight into our soma through somatic movement exercises. Somatic work in movement-based therapy helps individuals to integrate and recognize this connection and see where and how affect is burying itself in the body by examining and manipulating one's kinesthesia. Joan Chodorow (1991) explains that the idiosyncratic movements can express the unknown or repressed sides of the personality. That they are idiosyncratic harks back to my claim that improvisational practices of inquiry are subjective since these expressive movements of the inscape cannot fully be made sense of by witnesses. As can be expected through any form of therapy, but what can be surprising about movement therapy, is the extent to which repressed emotions accessed through kinesthetic-affective movements can be fully overwhelming and consuming in ways that verbal dialogue cannot be. This points to the unassailable ability of improvisational movement to access embodied knowledge to communicate qua movement-knowings.

According to Chodorow, motor imagination operates as a form of active imagination, which can provide one with more direct access body memory. Motor imagination, as active imagination, consists of rhythms in the body and relationships to the body that can help one develop a self-directed consciousness. This is why dance is considered to be a natural conduit to affect and corresponding body memories. Chodorow gives a detailed explanation of the somatic process and how active motor imagination pulls out body memory content:

> In the early stages of the work, the mover may remember very little. But gradually, the mover learns to allow the movement to unfold and at the same time pays attention to what the body is doing. The mover also develops a more differentiated sense of where the movement comes from, i.e. the

> inner world of bodily felt sensations and images. When the mover is aware of both the expressive action and its sensory-imaginal source, s/he is likely to be conscious of the emotion or emotional tone as well…for example, the emotions may by strongly felt on a visceral level, but without awareness of the images that would give it meaning. On the other hand, there are movers who experience the unconscious as an inner landscape rich with mythic images, yet their relationship to it may be curiously detached, as if it had nothing to do with them. In time, active imagination in movement tends to develop in us a more differentiated, balanced relationship to both the sensory and imaginal aspects of our emotional life (114) …. There are also times when the images themselves seem to want to be embodied, as if the image could make itself better known by entering the body of the mover. When this occurs, the experience shifts from *dancing with* a particular image to allowing oneself to *be danced by it*". (126)[8]

Active/motor imagination surfaces past emotional experiences through (kin)aesthetic movement. Moving via dance can prompt unbidden memories to manifest as particularly vivid images or reexperienced physical sensations. These visual and kinesthetic re-experiences are displayed on, and through, the body as it dances through movement and takes moments of somatic pause, especially if the revisiting of the emotional experience is being resisted.

As I mentioned above, somatic movement is enhanced through the involvement of a witness. In the case of dance movement therapy, this may seem a bit counterintuitive since somatic movement is largely subjective and, therefore, is unlikely to be read well by others. However, part of our soma is constituted by our existence as individuals in relation, in society, in culture—our inherent intersubjectivity. This is why a witness—the therapist, in the case of dance movement therapy—is almost an essential aspect of this kind of self-reflexive, self-reflective work. One might wonder, then, how a witness can contribute to the process. Diana Fischman (2016) rightly informs us that successful intersubjective assistance is made possible through the building of kinesthetic empathy within the relationship—often through practices of mirroring and affective attunement. Dance movement therapy is an inherently relational practice and so the therapist

[8] This point is very important to my method because it coheres with one of my own somatic exercises that I used in the experiment but that I borrowed and adjusted from other somatic exercise books.

becomes equally involved—as a partner—in the intersubjectivity of the experience that is rooted in the individual practitioner's body (34). Fischman explains that communication between witness and mover is established through all of the available sensory motor channels that comprise our soma. In a sense, then, and as I will explicate later, this communication that emerges through the establishment of kinesthetic empathy—an empathy in the flesh—in the relation exhibits a kind of intermingling of both person's somas, given the porosity of our soma qua affective-sensory intersubjectivity. Fischman identifies this porous merging of the somas as a transitional space, which is an area developed by the self and an Other, of the inner and outer worlds, that allows for primary creativity and the sensing of self-existence (40).

Advantageous to my purpose here is Fischman's argument that DMT itself is an enactive procedure. This is because, on her account, the therapeutic process is an affective-cognitive-creative experience demonstrative of a shared adventure (47). However, her claims regarding DMT are more of a gesture than an argument. But she is not wrong.

One way of positioning somatic work within the enactivist camp is by aligning the practices and techniques of somatic work with a model of Affective Learning Design for expertise development, which facilitates the purposeful generation of embodied virtues. This model indicates that how a performer feels greatly influences how they will perform. This shifts somatics, as a process, away from traditional somatic theory, á la Gendlin and Damasio, into a more radical enactive model that assigns emotions more intentionality in affective evaluation of external stimuli; or rather, ALD attributes intentionality to emotions since they are so extensively determinative of our world-directed activity. Accordingly, intentionality becomes constitutive of somatic movement because "*[s]omatic intention is directed intention* involving conscious use" (Fraleigh 2015a, xxii).

Radical Enactivist theorist Daniel Hutto et al. (2019) states:

> In contrast we think that emotions can be targeted and thus world-involving, and that emotional appraisals take the form of embodied evaluations that aren't dumb at all, and that it is possible to make sense of these ideas even if emotional intentionality and appraisal are not constituted by any kind of contentful attitude (31) …Enactive approaches go even deeper into the fleshy, bloody body, further than the skin, it goes into the world and back. (34)

While Hutto et al. are largely concerned with the affective evaluation of expert athletes in performance, it should not go unnoticed that practitioners of somatic work develop a very unique and grueling kind of expertise that can require a lifetime of intense work to achieve. From a DMT perspective, attributing this level of intentionality to emotions helps show how the kinesthetic-affective sensations unleashed during somatic work can be purposefully aroused for reflection. Hutto et al. call this a "full-bodied appraisal" (ibid.), which I believe best exemplifies the integral purpose of somatic work. And somatic skills can unironically be said to take place in the fleshiest, bloodiest spaces of all.

The Priority of Somatics

Susan Aposhyan once said that "by watching the movement of the body, we can see the movement of the mind" (Brodie and Lobel 2012, 128). This is a key insight as to why I believe that improvisational dance can be used as a mode of embodied critical inquiry; dance is a means through which we can read bodied language and actually translate back into static language—though the second step is not necessary. In this section, I lay out the distinctive role of somatic work in improvisational dance. For not all improv work relies on, or draws from, somatic work. Lots of improvisational work is used for creative exploration using difficult or silly prompts in order to help movers think outside of the bodied-box, as it were. Improv can be a warmup, it can be a challenge, and it can be spontaneously expressive in relation to a song that's playing. But somatic practices are considered a radical practice (Braude and Shulman 2015, 124) because they are used with great purpose in improvisational exercises where the objective for the improvisational movement is to connect with the movement-knowings. I'm inclined to maintain that some of the more free-form and fun-loving improvisational practices lack the necessary attributes to serve a role in an embodied critical inquiry. This is not particularly controversial since what makes an embodied inquiry critical is its purposeful directness.

Fraleigh (2015a, 2015d) highlights some of the specific functions of somatic work in choreographing processes that reveal why I hold that somatic practices are necessary to an embodied critical inquiry. The most substantial function is the role of freedom and choice in, and for, the movement.[9] While this might seem contrary to my above distinction between somatic-based vs non-somatic improv, this actually holds. This is because non-somatic improv, when used as a creativity device, can sometimes feel very contrived and leading in relation to its prompt. Because somatic work doesn't necessarily begin with a prompt, it leaves the body freer to make more choices for movement that is consistent with the body's logic. There is a sense, then, in which true creativity is spawned because the body is directed inwards towards the kinesthetic-affective; it is then directed by the intentionality of emotion wherever the body needs to go to articulate the perceived embodied knowledge. The body is not nudged in any one specific direction or in any one specific manner; the body goes to the experience that awaits it. Fraleigh eloquently captures this experience when she says: "Before the conscious mind connects the dots, we have already embodied what we value, experientially" (2015a, 30). She develops this idea further when she says:

> Clearing is also apparent through intuitive dance processes. Participants often describe what has become clear to them through the dance, what has improved, or what they have learned. (2015d, 70)

And Midgelow is right when she asserts that somatic improv is a critically embodied thought. The freedom of the body to access its movement-knowings and follow its body logic is precisely why somatic-based improvisation is an active mode of thinking and questioning.

From an enactivist perspective, this supposition of enhanced freedom and choice in somatic-based improvisational movement can be explained in terms of the "as-if" affordances that open up when one is led by their movement-knowings. Somatic movement is used to access embodied affects related to lived experience, but somatic movement also colors the affect found during movement. Improvisational movement led by somatic priming is a kind of embodied dialogue with embodied knowledge; it is a doing and being of, and from, that place of lived experience. As such, the movements of improvisational dance generate their own affect that then intermingles with the found affect in a way that pushes out a new affect, which is the content of the movement-knowing in that moment; this

[9] There is a sense in which Brian Massumi's account of the virtual could be paired well with an account of one's relationship to movement in somatic work. See: *Parables for the Virtual: Movement, Affect, and Sensation*. Durham: Duke University Press, 2002.

content is the content of the embodied critical inquiry. The as-if affordances are the possibilities of choice that open up to the mover in just the same way that conversations—with self or other—opens many lines of explorations. Choreographer Ami Shulman points to the importance of somatic-driven improvisation as-ifs:

> What becomes available to me in my movement completely changed simply by the difference of the angle of my head and viewpoint. The Feldenkrais [somatic work] perspective calls this a shift of habitual organization, which also shifts one's self-image, giving rise to previously unexplored possibilities (Braude and Shulman 2018, 184) … At some point in the improvisation, I do a movement that intrigues me, and I instinctively know that I want to make an exercise leading out from this point of entry. The movement affords what I can do with it (194) … This is an invitation to possibility in movement through embodied thought. (195)

Though Shulman is conceiving of these possibilities in terms of a "what if", it is clear that this approach to, and perception of, choice in somatic-inspired improv is paradigmatic of the enactive as-if. As Albright says, improv is a "what if" and not a "what is"; that is, she argues that improv is a practice of dwelling in possibility (Albright 2019, 25). Freedom always exists within contexts and constraints—freedom and choice in movement is no different. The somatic priming of improvisational exploration engenders the appropriate constraints for freedom of embodied pre-reflection that will then make reflection more intelligible. Unconstrained or poorly constrained movement, much like unconstrained anything, makes sense to no one and can be made sense by no one, including oneself. In an enactive approach, the basic constraints are generated by the conditions afforded through the body-brain-environment dynamical system. And because the body-brain system is directly attuned by affect, one's encounters with, and manipulation of, affect through movement in one's environment provides ample constraints conducive to sensical affordance exploration indicative of freedom.

One important aspect of the embodied critical inquiry method that I discuss in the method chapter is that I encourage the use of visual imagery in the somatic priming phase but discourage it in the immediately impending improvisational work; I strongly promote the use of imagination, but I strongly dissuade the mover from resorting to the use of visual images.

This is a small point but one that raises some interesting questions given that DMT often incorporates visual images (mental or physical) in their somatic-based dance work. One will be led to ask: how does one fully engage their imagination during movement without also relying on visual images? The distinction I make is related to a distinction between the kind of mental representations put forward by traditional cognitive science and a different version of mental imagery conceptualized in radical embodied cognition schemas. From a traditional standpoint, mental images are construed as static snapshots taken by our brain and then stored for later use—they are mental representations in the most customary sense. But enactivism is a non-representionalist framework and so it is important to understand just what is going on when one uses their imagination from a non-representationalist perspective.

Our imagination is comprised of very dynamic cognitive processes that are necessarily embodied because it is integrated with our perception. But even more, imagining is embodied because the things we imagine emerge from our motor imagery—images, then, are motoric bodily processes, not representations (Gallagher and Rucińska 2021, 2). Ultimately, given that embodied cognition is engendered by, and through, the body-brain-environment, our mental images are an admixture of movement, affect, and know-how knowledge (9). Roughly speaking, mental images are equal to the motor processes involved in imagining and are not, in any sense, a product of imagining—they are not snapshot images that are stored in the mind as representations. We engage in processes of imagining when we rehearse the germane movements or actions; movements constitute imaginings (10). Gallagher and Rucińska explain that "[my] neurons, and my body, go into certain states when I engage in such processes, and these are material processes that we can explain and sometimes measure" (ibid.). The kinds of movement-based rehearsals—the activities of imagination—that generate mental images are practices of marking (think cheerleading) and pretend play. When viewed in this way, we learn that our imaginings for movement and action are constrained, even in our imagination, by possible affordances of the body and the situation. This analysis of imagining is fully consistent with my above claim that freedom of choice in somatic and improvisational movement is achieved within the constraints of as-if affordances. It also reveals that the body can imagine and be imaginative in movement outside of visual representations.

Bibliography

Albright, Ann Cooper. "Life Practices". In *Oxford Handbook of Improvisation in Dance*. Midgelow, Vida (ed). New York: Oxford University Press, 2019: 25–36.

Bloom, Lynne Anne and L. Tarin Chaplin. *The Intimate Act of Choreography*. Pittsburg. University of Pittsburg Press, 1982.

Braude, Hillel and Ami Shulman. "Radical Somatics". In *Moving Consciously: Somatic Transformations through Dance, Yoga, and Touch*. Fraleigh, Sondra (ed). Urbana: University of Illinois Press, 2015: 124–134.

Brodie, Julie and Elin Lobel. *Dance and Somatics: Mind-Body Principles of Teaching and Performance*. Jefferson: McFarland and Company, Inc. Publishers, 2012.

Buttingsrud, Camille. "Bodies in Skilled Performance: How Dancers Reflect through the Living Body". *Synthese* (2021): 1–20.

Chaiklin, Sharon. "We Dance from the Moment Our Feet Touch the Earth". In *The Art and Science of Dance/Movement Therapy: Life is Dance*. Chaiklin, Sharon and Hilda Wengrower (eds.). New York: Routledge Press, 2016: 3–11.

Chodorow, Joan. *Dance Therapy and Depth Psychology: The Moving Imagination*. New York: Routledge Press, 1991.

Eddy, Martha. *Mindful Movement: The Evolution of the Somatic Arts and Conscious Action*. Chicago: Intellect, 2017.

Fischman, Diana. "Therepeutic Relationships and Kinesthetic Empathy". In *The Art and Science of Dance/Movement Therapy: Life is Dance*. Chaiklin, Sharon and Hilda Wengrower (eds.). New York: Routledge Press, 2016: 33–52.

Fraleigh, Sondra Horton. "Prologue on Somatic Contexts". In *Moving Consciously: Somatic Transformations through Dance, Yoga, and Touch*. Fraleigh, Sondra (ed). Urbana: University of Illinois Press, 2015a: xiii–xxxi.

———. "Why Consciousness Matters". In *Moving Consciously: Somatic Transformations through Dance, Yoga, and Touch*. Fraleigh, Sondra (ed). Urbana: University of Illinois Press, 2015b: 3–23.

———. "Dancing Becomes Walking". In *Moving Consciously: Somatic Transformations through Dance, Yoga, and Touch*. Fraleigh, Sondra (ed). Urbana: University of Illinois Press, 2015d: 50–71.

———. "A Philosophy of the Improvisational Body". In *Oxford Handbook of Improvisation in Dance*. Midgelow, Vida (ed). New York, 2019: 65–88.

Gallagher, Shaun and Zuzanna Rucińska. "Prospecting Performance: Rehearsal and the Nature of Imagination". *Synthese*, 2021: 1–19.

Hutto, Daniel, Michael Kirchhoff, and Ian Renshw. "Emotions on the Playing Field". In *Handbook of Embodied Cognition and Sport Psychology*. Cappuccio, Massimiliano (ed.). Cambridge: MIT Press, 2019: 23–46.

Lavender, Larry. "Facilitating the Choreographic Process". In *Contemporary Choreography: A Critical Reader*. Butterworth, Jo and Liesbeth Wildschut (eds.). New York: Routledge Press, 2009: 71–89.

Midgelow, Vida. "Some Fleshy Thinking: Improvisation, Experience, and Perception". In *The Oxford Handbook of Dance and Theater.* George-Graves, Nadine, ed. New York: Oxford University Press, 2015: 109–122.

Protopapa, Efrosini. "Choreography As Philosophy, or Exercising Thought in Performance". In *Thinking Through Dance: The Philosophy of Dance Performance and Practices.* Bunker, Jenny, Anna Pakes, and Bonnie Rowell (eds.) Hampshire: Dance Books, Ltd., 2013: 273–290.

Rothfield, Philipa. "Chance Encounters, Nietzschean Philosophy, and the Question of Improvisation". In *Oxford Handbook of Improvisation in Dance.* Midgelow, Vida, ed. New York, 2019: 89–102.

Welch, Shay. *The Phenomenology of a Performative Knowledge System: Dancing with Native American Epistemology.* New York: Palgrave Macmillan 2019.

Williamson, Amanda. "Falling in Love with Language". In *Back to the Dance Itself: Phenomenologies of the Body in Performance.* Fraleigh, Sondra. Urbana: University of Illinois Press, 2018: 78–98.

CHAPTER 5

Embodied Critical Inquiry: A Theory

By watching movement of the body, we can see the movement of the mind.
—Susan Aposhyan

The goal of this chapter is to lay out the underlying theory behind an embodied critical inquiry. Throughout the book, I have addressed smaller, interrelated notions that contribute to the intelligibility of this theory and its implementation. I have also included extensive surveys of the various interrelated disciplines to close gaps that obtain in one that can be filled by the others—an issue that may be a looming mystery in one can be demystified when seen from the perspective of another. Because there are many interconnecting components, and because this project vacillates constantly between the theoretical and the contextual, I chose to divvy up substantial pieces for up-close analysis. In this chapter, I braid together the wild spralling of ideas regarding the cognitive capacities of the moving body. For this reason, there will be some repetition of ideas and descriptions from earlier chapters, but I try to keep this repetition to a minimum. Additionally, there may seem to be some remaining theoretical gaps, but this is one of the difficulties of an interdisciplinary applied philosophy. Fortunately, some of the points that I do not clarify here are made more evident in the methodology chapter, since the theory makes most sense in its application.

S. Welch, *Choreography as Embodied Critical Inquiry*,
https://doi.org/10.1007/978-3-030-93495-8_5

In the introduction, I assert that my method for an embodied critical inquiry follows Sheets-Johnstone's TIM model, which has been adopted and adapted within various embodied cognition schemas. For this reason, in "The Epistemology of Thinking-in-Movement" section, I provide a more detailed exposition of Sheets-Johnstone to explicate how, and why, the TIM model is a discrete epistemology—it is not a propositional epistemology and it is not merely a procedural epistemology; it is a moving epistemology. While I have shown throughout that dance is a venue for better understanding embodied cognition, I have not directly addressed its epistemology. This task is important because it is possible for an unarticulated, assumed epistemology to work against the model of embodied cognition. Additionally, the specific kind of epistemology that grounds thinking-in-movement is important for explaining how, from a cognition perspective, dance can actually *inquire into* conceptual problems. In "Epistemology and Dancing Metaphors" section, I take a brief interlude to address embodied metaphors more thoroughly since these are active throughout the chapter. While I have already touched on embodied metaphors, I have not yet explained how they operate within dancing specifically. Additionally, while setting out how embodied metaphors function as dancing metaphors, I address their epistemological powers and effects. In doing so, I show how embodied metaphors smoothly fold into the thinking-in-movement epistemology.

In "Inquiring-in-Movement" section, I take on the question of how movement and dance are an unmediated means through which to query. This section is the core theory of the method. While I try to avoid going into methodological exercises and practices, I do discuss how somatic work and embodied metaphors do the heavy lifting involved in embodied questioning. This helps the reader gain a firmer grasp on what, exactly, I'm going on about and how this actually plays out. I have been referring to this process as an embodied critical inquiry but, here—since I am focusing on a dance context—I identify the method as an inquiry-in-movement to show its pairing with Sheets-Johnstone's dance context for the TIM frame. In the "Embodied Dialogues" section I explicate how thinking-in-movement and inquiry-in-movement can be successfully executed between movement partners. This process of an embodied dialogue relies on an account of social cognition and intersubjective participatory sense-making. These both come out of the enactivist framework but here I only discuss enough of the material to provide a grounding for partner work.

The Epistemology of Thinking-in-Movement

Sheets-Johnstone argues that improvisation is the quintessential process through which we can recognize thinking-in-movement (1999, 484). One reason speaks to a version of the implicit distinction I purported to inhere in Langer's work: there is a distinction in meaning and uptake between improvisation practices and performed choreography as presentation.[1] I suggest that the meaning of improvisational dance is not virtual because the dance is not being repeated, and Sheets-Johnstone makes a similar claim. And because improvisational movement is largely spontaneous, because it is a process and a becoming, it instantiates unmediated thinking in action. She rightly espouses that improvisation is a process of creation and that this process of creation is the dance itself. Thinking-in-movement is not a thinking about movement. The thinking is in and through the movement because we are a body-mind comprised of a soma.[2] Therefore, we can regard thinking-in-movement, especially improvisation, as evidencing the body-mind as truly non-dualist and non-representationalist. As such, it best aligns with an enactivist framework. Another reason thinking-in-movement—and accordingly, movement-knowings and curiousing—belong to the enactive embodied cognition camp is because of this constant becoming that characterizes improvisation. Enactivism is defined, in part, by dynamical systems. On Sheets-Johnstone's account, movement creation is not something done; it is always a dynamic process that cannot be divided into static pieces. The meaning that imbues and incites it is emergent from this dynamical system.

Given the incessant processual nature of creative cognition in both enactivism and improvisational praxis, an embodied critical inquiry must be grounded in a procedural knowledge framework. While thinking-in-movement is best characterized as a procedural knowledge framework, it is important to emphasize the extent to which choreographing, as such, is an epistemology in itself qua moving epistemology. What I mean by this is that thinking-in-movement is its own unique frame of knowledge. It has a distinctive way of knowing, coming to know, having knowledge, sharing knowledge, mining knowledge, storing knowledge, and understanding. The epistemological frame of thinking-in-movement has the processes and conditions for knowledge that stands apart from any other epistemology—though

[1] I want to note that improv can (1) also be performed in its own right and (2) even when it's not performed, it is performative.

[2] See also: Barbara Tversky. *Mind in Motion: How Action Shapes Thought*. New York: Basic Books, 2019.

it constitutes a large part of many Indigenous epistemologies. As Karen Barbour so eloquently announces: "We dance epistemologies in just such ways" (Barbour 2018, 246). The content of this epistemology is the phenomenological, subjective experience articulated through the moving body's knowledge and skills; this embodied sensory sensate sensing collective—this knowing-moving-bodying experience—is the "potentiality for thought, intuition, and imagination" (Grove 2005, 44). This is because thinking-in-movement is all from the inside (Sheets-Johnstone 1999, 495); choreography is, as one dancer puts it, a memoir (Kim 2016, 52). That is, there is no one for whom it is intended and there is no one for whom it is crafted; it is pure moving subjectivity. Or to rephrase, it is purely by and for the subject from the subject's position for the subject's own intention. While much choreography ends up as a presentation for others, the solo choreographing process is simultaneously prereflective and reflective, alternating between the implicit and explicit, but always from the inside out and always from the phenomenological first-person perspective in movement.[3]

The ability to engage in a critical mode of inquiry is one of our most invaluable epistemological skills. The epistemological skill set that comprises critical inquiry includes different means through which to enact and direct our creative cognitive skills to gather and disseminate knowledge successfully to satisfy some goal. Traditional epistemological accounts of critical inquiry center skills that relate to propositional forms of knowledge. But as I mentioned in Chap. 2, the epistemological approach that my theory turns on diverges from mainstream frames. So when one attempts to outline a background epistemology for an embodied cognition schema, it is imperative to explicate how, and why, the two differing epistemological models of critical inquiry are divorced from one another. There are three main distinctions between traditional critical inquiry and embodied critical inquiry: one utilizes propositional epistemological skills and the other employs procedural epistemological skills related directly to movement and motion, one seeks out information external to the body and the other seeks out information internal to the body, and one is abstract whereas the other is phenomenological. I have addressed some of these already, particularly as it concerns embodied knowledge and phenomenology. Yet these two coalesce in a distinguishing manner that effects the distinction between them.[4]

[3] One can conceive of a collective as choreographer and in this sense the "we" of the collective is still a form of pure subjectivity, as the dancers understand it.

[4] See: Katherine Hawley. "Success and Knowledge-How". *American Philosophical Quarterly* 40, no. 1 (2003): 19–31; Charles Wallis. "Consciousness, Context, and Know-How". *Synthese* 160 (2008): 123–153.

Katherine Hawley (2003) provides an effective account of procedural knowledge that I take as an exemplar for the epistemology behind TIM. She argues that one knows-how if she succeeds in her performance of the action and has a warranted belief regarding her ability to know how to succeed. In this sense, knowing-how is a form of practical knowledge since it requires an action about which we must have methodological knowledge. This methodological knowledge, which she describes as "received wisdom" (20) gestures towards the relationship between procedural knowledge and propositional knowledge. However, we need not be experts in methods on how to succeed in performing X to have a warranted belief that we can and will. For Hawley, S knows how to X under circumstances C but failure to succeed under extenuating conditions does not invalidate one's procedural knowledge about X. Additionally, it is possible to have procedural knowledge about things which we are not able to perform or, further, which we have not attempted because it is possible for us to learn methods for successful performance through practices of paying attention. This means that performance is the crux of knowing-how. For example, someone who is an expert dancer can surely give a hand to an ice skater who's trying to develop a routine.[5] Similarly, as Hawley explains, success under all counterfactual conditions isn't sufficient for knowing-how since one will surely fail if they're dancing on a sprained ankle. What is necessary, though, is that successful performance would ensue under certain context-dependent and context-relevant counterfactual circumstances. My friends who race cars and do motorcycles stunts might scoff at someone who claims to *know how* to drive.[6] So context *and* context description are crucial to evaluating the accuracy of one's claims regarding their procedural knowledge.

[5] She does make an odd claim that once you are no longer able to do an X that you were able to do before that that knowledge is then rendered useless. Her example involves a child who was in an accident. Before, she knew how to ride a bike with two legs but now she only has one. She rightly explains that the loss of one leg does not invalidate her knowing how to ride a bike with two legs but it does render that knowledge useless. I find this odd because some of the most amazing teachers are those who had to stop performing and start teaching because an injury prevented them from ever performing again. So I think her construal of "useless" is overstated since if that girl wanted to teach her friend who has two legs how to ride a bike then she could use her procedural knowledge of doing so to guide her technique.

[6] I must admit that after writing up this example, I went back to her article to do some extra reviewing and realized that she, too, uses driving as an example. So I just want to note that I did not hijack her example, it just turns out that driving seems to be a really obvious example for this point.

To better understand this, it is helpful to start with what Max Van Manen (2014) calls the Phenomenology of Practice and demonstrate how this grounds an embodied moving epistemology of critical inquiry. Van Manen states that a phenomenology of practice is a hermeneutical practice centered on a distinctively phenomenological kind of research. This form of research also involves writing that prepares one for the activity or allows them to reflect on it (15). Focusing on both performance and writing helps one discover, and create, the meaning within, and of, the research. Van Manen poignantly captures the significance of phenomenological experience to research in a way that also points towards the choiceness of an embodied cognition model:

> Meaning is not something that can just be scooped up from the spoils and layers of debris of daily living. Meaning is already implicated in the mystery of prereflective reflection of seeing, hearing, touching, being touched, and being-in-touch with the world, and the enigma of reflecting on the phenomenality of all this. New experiences may grant us unsuspected encounters with significances that we did not know before. Thoughtful reflections may bring ancient and novel sights and insights into perspectival view. The phenomenological pathos is the loving project of bringing all the living of life to meaningful expression through the imageries and language of phenomenological writing, composing, and expressing (18)…Phenomenological method is driven by a pathos: being *swept up* in a spell of *wonder* about phenomena as they appear, show, present, or give themselves to us. (26, italics mine)

The embodied nature of the phenomenology of practice is decidedly apparent here. Another sense in which his schema is embodied is in his assertion that once one starts to reflect through language, the 'what it is' has already become the 'what it was' and the experience becomes objectified. The reflection is on the embodied subject who is/was experiencing. This recognition of the inevitable passing of subjectivity through reflection reveals the inevitability of inquiries that are not embodied to be necessarily mediated. Notably, while the purpose of a hermeneutics of phenomenology is to reflect on the practice to make the experience of the activity intelligible, Van Manen gestures towards the method's primary function of questioning; he calls it a "wondering pathos" that is necessarily based on openness (27). His analysis of a phenomenology of practice coheres with my conception of critical inquiry constituted by an analysis of movement-driven curiosity that is imbued with wonder and everlasting creativity. He claims that the inherent creativity within phenomenological practices situates it somewhere between order and disorder, analysis and imagination (72). But because the requisite degree of creativity is so intense, it is by nature radically dynamic.

This account of a practice-based mode of research concerned with experiential reflection through inquiry is an appropriate foundation for anchoring an embodied critical inquiry in a procedural epistemology frame. The only inadequacy of this model lies in the prioritization of written reflective practices, which pushes it back towards the propositional side of the epistemological spectrum. That is, there is too much writing-hermeneutics and not enough moving-hermeneutics. But this is not to say that writing practices are not germane to an embodied critical inquiry; it is just that the body and its knowledge exceeds and overflows the capacity of language to account for experience. An embodied critical inquiry does not preclude a role for propositional knowledge, but it does prioritize a moving procedural knowledge, which makes sense given the priority of somatics in the method. Writing reflections of various sorts via worksheets is an essential component of an embodied critical inquiry (see methodology chapter). However, Practice as Research models that are becoming established are ideal forms of phenomenology of practice because they invert the role of performance and written reflection.[7] Understanding just is the intersection of "bodily adventure and language home" (Todres 2007, 24). Practice as research can be considered as almost entirely a procedural phenomenology of practice that includes reflection, but this reflection is the movement—the TIM framework. Even Midgelow recognizes the connection between somatic-based improvisation and Van Manen's work. She explains:

> Improvisational knowledge is pathic to the extent that the act of practice depends on somatic sensibilities and relational perceptiveness, alongside other aspects of knowledge, that are rigorously pursued and that are in part pre-theoretic and pre-linguistic [(Van Manen 2007)] … it is significant that this "thingness" in somatically informed improvisation is focused toward attitudes of attending and processes of opening. (Midgelow 2018, 61, 64)

[7] See also: Barbour (2011); John Sutton. "Moving and Thinking Together in Dance". In *Thinking in Four Dimensions: Creativity and Cognition in Contemporary Dance*. Robin Grove, Catherin Stevens, and Shirley McKechnie (eds.). Melbourne: Melbourne University Press, 2005: 50–56; Wengrower (2016).

An embodied critical inquiry is most certainly a form of hermeneutics; but it is an embodied hermeneutics that is specifically a phenomenology in action.

Accordingly, another way of conceiving of an embodied critical inquiry as having its own moving epistemology frame is in the way it can be conceptualized as a form of process philosophy.[8] A process philosophy focuses on the dynamics of interrelatedness in life and in experience. But instead of being concerned with the static relation, it is interested in the processes that spawn and sustain those relations. Yet again, the principal operation here is the incessant nature of creativity and becoming that propels the dynamics of life onward. In her attempt to connect process philosophy with improvisation and phenomenology, Midgelow further argues that:

> Expanding this sensibility improvisationally, we learn to understand how perception is in constant flow and how to research in, through, and about that which is in motion. They ask us to open a space of possibility, making us more critically and somatically aware such that the "how" of being able to notice is transformed and our bodily knowing becomes more acute. (76)

Thinking-in-movement that motivates Practice as Research is a process philosophy and process philosophy is tied to a procedural epistemology. Sense-making qua movement, while subjective, is not a personal venture. The lived—moving, sensing—body serves as the authenticating *procedure* for, and *process* of, sense and knowledge (Todres 2007, 31).[9] Over and over the explicit role for embodied knowledge manifests in movement and a distinct epistemology shines through.

Na-ye Kim and Anna Pakes attribute a distinctive epistemological status to choreography vis-à-vis the procedural know-how internal to phronesis. Kim seems to pull choreography and phronesis together indirectly by targeting the creative cognition capacities of choreography; on her account, choreography is a knowing practice because the activity of creating requires critical reasoning (43). While Pakes paints the phronetic power of choreography similarly in terms of the creative cognition inherent in practical reasoning, she rightly espouses the fact that practices of choreography are a knowing in their own right (Pakes 2009, 10). That choreography is a procedural, practical wisdom kind of epistemology is rather

[8] Two of the most notable process philosophers are Alfred Whitehead and Henri Bergson.
[9] See also: Grove (2005) and Grove and McKechnie (2005).

straightforward. That it is an epistemology in its own right might be more contentious. But Pakes hits very close to the precise reason as to why choreography is a discrete epistemology; or rather, Pakes nails an aspect of choreographing that I think captures how an embodied critical inquiry operates as movement that can be "read" in a way to divulge embodied knowledge: the placement of conscious intention. Pakes provides a teleological account of intentions in choreographing. She argues that the intentions thrusting creative movement during the choreographing process can be seen after the process takes place. The reasoning behind the need for certain movement is backward-looking, in that it is determined once the process is complete and is then used to justify the creative process in relation to its ends (14). When choreography is Practice-as-Research, the intention behind the practice-research can always shift and can often be discovered in it (17). It is the ability to discover the intention behind the creative movement—the curiousing—that indemnifies improvisation's ability to function specifically as inquiry, rather than as purely narrative.

Our tacit and implicit ways of knowing, such as embodied intuition and body memory, are direct and continuous performances by our bodies of peculiar kinds of moving know-hows. Understanding the progression from deep embodied meaning-making to implicit and explicit knowledge is crucial to this project; how one gets from deep embodied knowledge at the cognitive level to an analysis of experiential, praxis-based procedural knowledge is at the core of dance phenomenology. Dancer Monique Mojica (2009) explicates this idea by saying that:

> ...our bodies are our libraries—fully references in memory, an endless resource, a giant database of stories. Some we lived, some were passed on, some dreamt, some forgotten, some we are unaware of, dormant, awaiting the key that will release them. (97)

She relies on praxes of improvisation as a method of "mining" her body for "organic texts" to motivate her choreographic storytelling. This embodied epistemological practice of know-how that comes from and through movement is what I mean when I say that improvisation can access movement-knowings subjectively and draw them out for communicative purposes.

Epistemology and Dancing Metaphors

According to embodied metaphor accounts, our situatedness opens but constrains the conditions for action through which we generate meaning. Meaning constructed through afforded action extends metaphorically in the form of image schemas in our conceptual mappings (Johnson 1987, 2007). An image schema develops when our sensorimotor experiences track repeated patters and relations and so frames our "meaningful" experiences. Examples of embodied perceptions that engender meaning include verticality, twisted, circular, toward, away from, into and out of, sharp, hot, shape, and rush. These sensorimotor experiences that result from our situatedness qua the body-brain-environment dynamic generate our source domains. We then extend the meaning entailed by primary metaphors to analogous conceptual ideas, which are target domains. As one can see, these basic image schemas are surreptitious in daily life, but are exceedingly visible in dance. Dance is not too far away from our basic sensorimotor habits and contingencies; they are just performed with more panache. Dance and cognition scholar Jaana Parviainen explains:

> Because the body we live has an 'intelligence' of its own, it can tell us, sometimes very precisely, what it is we need from the present lived moment of our historical situation. There is a *givenness* concerning embodiment and participation in nature, cultural and social fields. (1998, 30)

This is one of the reasons that embodied cultural and social analyses include, or are reliant on, accounts of body memory.[10]

Midgelow rightly asserts that metaphoric images effect kinetic knowledge (Midgelow 2018, 73). An example of a conceptual metaphor that captures kinetic knowledge—and, accordingly, an embodied metaphor—is "I am caught up in my feelings". This example actually counts as a blended metaphor since it is comprised of two operative image schemas: up and in—there is a container and there is height. The container here is represented by "caught" and the height is represented by "up". Caught implied being trapped within a container. Up, which is a height, can operate in a few ways. It may signify something positive, such as "I am feeling up", or it may signify something lofty, such as "My head is up in the

[10] See also: Michela Summa. "Body Memory and the Genesis of Meaning". In *Body Memory, Metaphor, and Movement.* Koch, Sabine, Thomas Fuchs, Michela Summa, and Cornelia Müler (eds.). Philadelphia: John Benjamins Pub. Co., 2013: 23–41.

clouds"; but pertinent to the example here, height can indicate a danger, such as being up in a tree or up on a ledge. When you are "caught up in your feelings", you are experiencing a terrifying sense of entrapment that suggests a fear of falling. These conceptual metaphors, while taken for granted, are highly imaginative conceptual constructions that evolve from our embodied experience in the phenomenological world. Without our body's capacities to act—to move, perceive, manipulate, and engage—we would have no source from which to imaginatively draw ideas, induce, or infer. As I explicated in the previous chapter, imagination itself is a function of this embodiment at the deep level and therefore cannot spawn meaning and concepts on its own; imagination requires meaning from our sensorimotor patterns that ground our motor imagery—both of which contribute to the make-up of our kinetic knowledge.

For the purpose of choreography and embodied critical inquiry, performed bodied embodied metaphors work in the reverse of linguistic embodied metaphors. Performance philosopher Einav Katan-Schmidt (2016, 2017) powerfully explains how the cognitive experience of meaning and meaning-making of dance resides directly in physicality because dance is the manifestation of meaning qua movement.[11] Dance *articulates* meaning through the use of somatic precision in and through movement. Because dance does not represent meaning, but rather holds, shapes, and lives meaning, Katan-Schmid posits that dance must be danced for the ideas to be manifested and realized (2016, 16). It is through perceptual processes that dance both expands the cognitive capacities of the dancer and makes dancing cognitively intelligible to viewers. According to Katan-Schmid, the creative cognitive process of engendering meaning through movement exists at the point where sensuality meets intentionality. Fundamentally, active perception in dancing necessitates that dancers gaze into the sensual sources in, and of, their bodies and attempts to interpret understandings of those sources in relation to movement (50). She explains that since qualities of movement originate in embodied felt experience, dancers must intentionally pull from sensation and direct them into their purposefully constructed dancing metaphors (191).

From this, one can see that dancing enjoys its very own epistemology because it has the ability to be true (Fraleigh 1987; Welch 2019). Dance conveys truth so long as movements shaped in accord with movement-knowings proceed from embodied logics. When we go directly

[11] This discussion of embodied metaphors also appears in Welch (2019).

to the source of the body, embodied metaphors qua dancing metaphors become easier to follow. Dancing is a direct communicative action of these kinds of logical entailments that instantiate not only through the developed and selected movements, but also the transitions between them that constitute sequences. The movements are metaphorical ideas, the sequences are larger abstract narrative claims, and the transitions are the actual logical relations. The embodied kinetic logic requires that once I leap into the air and assume a shape to convey the idea, I must then—logically—find a narrative connection between the airborne metaphor and another on the ground. Embodied logic dictates that the leap plus the shape actually depict some related embodied experience. If I'm going for something narrative related to containment when I choreograph my aerial routine, my movements will necessarily require that I utilize a lot of wraps around my body or else it just doesn't make sense *logically*.

Embodied metaphors are particularly effective in dancing since movements can evoke abstract concepts and images. The imaginative yet rational structure of embodied metaphors function as enactive instructions through which to shift the perceptions of the dancer or viewer by directing their perception toward the phenomenological understanding of the related embodied experiences motivating the dancing movement pattern. I think that what can make them exceptionally communicative is an emphasis on precision in movement—"[t]he dancing body gives clues. It "talks" and people "listen" (Hanna 2015, 47). Katan-Schmid points to the significance of subtleties and intricacies like body texture. Such precise control over not only movement, but also body presentation, ensures a myriad of available embodied metaphors from which, and with, to work. I posit that dancing metaphors are epistemological because they blend both imaginary and actual contexts and conditions at the embodied level for both dancer and viewer. It is the melding of what is with what might be that is the operative cognitive logic that leads a viewer to see what knowledge the body is sharing through the embodied storytelling of the dancer.

Our embodied kinetic logics is constitutive of our cognitive unconscious and grounds our embodied implicit knowledge. Dancing, as a paradigmatic mode of thinking-in-movement, is an ideal medium for meaning creation and knowledge communication because it is centered on the juncture of movement-based logic/affordances and embodied imagination; dance facilitates the purposeful extension of embodied metaphors to other meaningful aspects of our experiences.[12] Thinking through movement involves a particular body logic and when exploring conceptually, choreographers merely have to follow the affects provided by the enactive affordances through the movements to their logical conclusions. Susan Foster cleverly depicts the body logic of choreographing. She enthusiastically pronounces:

> [Bodies] develop choreographies of signs through which they discourse: they run (or lurch, or bound, or feint, or meander…) from premise to conclusion; they turn (or pivot, or twist…) through the process of reasoning; they confer with (or rub up against, or bump into…) one another in narrating their own greater precision to our understanding of the reality of embodiment. (Foster 1996, xi)

When we access our movement-knowings through somatic-based improv, we become more consciously aware of the discursive content of that embodied knowledge because we can "read" the movements and identify the metaphors that surface from the cognitive unconscious and are displayed on, and through, the body; these are the embodied—or rather, dancing—metaphors. It is through teaching practices of manipulating embodied metaphors in choreographing processes for the purpose of narrative construction that I engage local arts communities in a form of public philosophy. While this idea is clearest when contemplating choreographed narratives, this truthing in, and through, dance also obtains when dancing is questioning.

Inquiring-in-Movement

In the previous section, I explained how dancing is a moving epistemology by explicating how meaning obtains within movement, how movement and metaphor are linguistically interdependent, and how movement meanings have a logical structure. All of this was to draw out the ways that dancing—specifically through dancing metaphors—can be communicative and true. I generally focused on narrative modes of choreography since these are the most straightforward, epistemologically speaking. But I also started with embodied metaphors and choreography to set a firm ground

[12] Just a reminder that I will be explaining how conceptual metaphor theory is not wholly inconsistent with an enactive approach to embodied cognition in the following chapter.

from which to help folx start to think about how all of this works in reverse. Once one understands how one can tell truths through movement, it becomes much easier to see how it is possible to ask oneself deep, meaningful, logical questions through movement. Through movement, the body dips into embodied knowledges through movement-knowings and serves up embodied metaphors that can then be interpreted for its embodied cognitive content that answers those questions. In this section, I focus on the role of thinking-in-movement in practices of critical thinking that are movement-based. This brings us back to the significance of somatics in choreographing and how it operates epistemologically in inquiry.

All movement involves thinking-in-movement and all thinking involves movement. An embodied critical inquiry is just one specific kind of thinking-in-movement. When thinking about movement-based activities, such as dance, martial arts, and sports, decision making is inherent in the performance of the action itself. Giovanni Pezzulo (2015) explains the epistemic operations of skilled performance as an action-perception loop. Our sensorimotor and affective sensations guide the sorts of actions we perform but the performance of those actions then effect new sensations that open up new content for decision making (21). These sensations and the possibilities they afford inform and structure action, which can then be exploited to direct our perception and learning (ibid.; Hanna 2015). To see the link between skill and sensorimotor control is to see the knowledge internal to performed movement. And, further, it is this link between performed movement and decision-making that makes dance suitable as an embodied medium for query.

One distinction I want to parse before moving on is the underlying, possibly indiscernible, distinction between how the questions derived during an embodied critical inquiry are at all different from the motivating question for that inquiry. One might presume that I have been using the notion of an embodied question as a substitute for the motivating question. If this were true, there wouldn't be too much controversy or too much that is all that interesting about an embodied critical inquiry. One might assume that the mover begins with a direct question and then engages in creative movement to see how the body relates to that question. In that case, the pre-posed question is a motivating question. But this is not the process I aim to evidence—though certainly this kind of process is both useful and effective for critical thinking and critical inquiry processes. What I am talking about is starting with a notion and then engaging a series of embodied movement exercises—mostly a combination of somatic and improv—to actually find what questions one even wants answers to within in the body by revealing them through the creative movement. This process is a bit harder for some to take seriously. Even the movement artists I worked with were initially highly skeptical about the potential for this process to succeed. It is important for the reader to keep at the fore just how literal I'm being when I talk about mining the body for questions using creative movement to access movement-knowings through an embodied curiousing.

Choreography clearly develops problem solving skills and aids in problem solving itself, but it also lends itself to conceptual inquiry. This is a choreographic capacity to grant access to the idea that one can engage in embodied critical inquiry conceptually. The very act of choreographing vis-à-vis sequencing, movement reformulation, and modulating dynamics directly involves conceptual decision-making (Hanna 2015). Movement need not be verbally motivated. Creative cognition through movement can be prompted by affective concepts provided through, or by, modalities that offer manifestations of felt sense, such as rhythms. The ability of ineffable cognitive motives or impulses undercuts the presumption that movement bearing cognitive content requires an input of representations and results in an output of representations (Stevens et al. 2003). In describing her experience with improv and somatic practices, Catherine Schaeffer imparts the following powerful testimony of what can be identified as a form of embodied conceptual inquiry:

> Shin somatics enhances the development of imaginative wisdom in performance. There is an ease with which artistic practices and life merge. Once a dancer trusts their inner knowing, they begin to move intuitively. This enables a dancer to become an agent of communication, aware of the entire process of dance itself. This conceptual gestalt changes the dancer's reality, from musculature and body physiology to a theoretical and visionary process where the brain is performing as the body speaks. When I transform into my creative state, a waterfall of instinctual movement flows out of my body. (Schaeffer 2015, 85)

What is going on here is that the body is intuitively accessing conceptual movement-knowings through somatic-based, improv-based inquiry. This is a form of curiousing that results from the body intuitively adhering to the kinetic bodily logos to delve into, and select, conceptual epistemological inclinations from the emergent embodied affordances afforded through the creative and exploratory movement process.

Contrary to what Sheets-Johnstone (1999) has argued, given the link between kinetic bodily logos and emergent affordances, embodied choices are both intentional and explicit. One does not arbitrarily pursue a movement even if one is moving intuitively. Intuitive movement is intentional in the thick sense; it is not accidental. This is not to say that "accidents" never happen, but when one is participating in somatic-based improv practices, it is more of a colloquial habit to identify a surprising spontaneous movement as an accident—these movements are found but chosen movements. If these "accidents" were not chosen, the mover would have circumvented the final position (of course, we all have real accidents). Moreover, I would disagree that such movement dynamics and pathways are not deliberated on. The presumption that expert "online" activity, such as dancing and sports playing, is automated and does not involve deliberation is spurious and fails to comprehend the sheer cognitive processing speed of expert decision-making. In the context of expert skill performance and online decision-making, Rucińska and Aggerholm (2019) call this kind of enactive process a creative responsiveness. That is, when one is able to make such "automatic" decisions, what they are doing is creatively responding to actualized affordances that become available for action. This idea seems less provocative when we consider that the emergent affordances upon which we decide, and from which we choose, are actually referred to as solicitations. If movers weren't making decisions in real time among a multitude of oncoming and enveloping sensory information, there would be significantly more injuries and bad calls. The intersection between kinetic body logos and emergent affordances is where online action-based embodied conceptual decision-making occurs—and this requires online action-based, movement-based inquiry with respect to the accounting and assessment of that onslaught of oncoming sensory cognitive content. Forward-looking, moving choice-making necessitates moving-inquiry. But she is right when she argues that going with the movement, and staying with the movement, means choosing to stay in relationship with that movement; and I would add, what that movement says.

One reason why movers might assent to the viability of an embodied critical inquiry more readily than others is because creative movers, especially trained movers, develop the distinguishing cognitive capacities of embodied critical thinking. These capacities and abilities expose them to ways that movement can lead one to learn about their relationships to ideas through their relationships to, and through, movement. Dancing, and other modes of creative movement, broaden our ways of understanding our world—and questioning our role in it—by broadening our kinesthetic capacities and knowledges. The reason why expanded sensory modalities and operations broaden our understanding of the world is because these modalities give us more exposure to divergent and plural affordances from which to make choices qua critical inquiry. It is not only that creative movement helps us better understand our place in the world; creative movement literally provides individuals more cognitive resources and angles from which to take the world, our place in it, and our relationships in it, into consideration.

An embodied critical inquiry is not motivated by a specific question since the question is what one is looking for. But a prompting concept is used because one needs a starting point. I have been discussing how all of these practices help one to question the world and one's place in it. But this is a very vague and almost unhelpful ways of describing the purpose of critical inquiry of any kind. Usually, we know what our topic is. This is also true of our philosophy assignments as much as it is true for artistic creation. Someone comes to you, or you come to yourself, and says: tell me about justice, tell me about beauty, tell me about trust or identity or freedom, tell me about loss. Someone says: hey, tell me about epistemic injustice and intersectional feminism. What then, does one do? Well, we have the concept. But then, if we're going to write that paper or we're going to create that piece for the show, we need to know what kinds of questions we have about that idea. So there's a sense in which our prompting concept is what we question about the world and then we have to further figure out what our place is in it. The paper, the piece, is the narrative of the world and our place in it—it is the concept and what we think about it.

But how in the world are we supposed to just say what we think about something without figuring out the questions we even care about in relation to the concept? If someone comes to you and tells you to make a piece about beauty, that is so overwhelming as to be ineffectual. To be able to make your piece, you have to find your relation. If there's a topic on beauty and there are five people, and each person goes through a preliminary process of critical inquiry before jumping into their narrative, there are going to be five very different narratives. There could be narratives on eating disorders, nature, death, fashion, and the color red. And each of these narratives will have come from each person's questions about the concept that helped them make sense of their relationship to that concept. Each asks about beauty in the world, but then each asks about their place in that world of beauty, and that is how we end up with so many wonderful, wonder-inspiring, awe-striking perspectives on existence.

Typically, people employ various kinds of games or tasks to help them with critical inquiry processes. One that automatically jumps out to me is brainstorming; this is taught to us very early on. We put the concept in the middle and then we make some kind web of related ideas around it. Or we free write until we hit something that sounds right. Sometimes we stare out the window purposefully yet mindlessly until something pops into our heads. And some people move until the body pushes out the right questions. It is a more intentional codified version of this process that I am thinking of when I am talking about an embodied critical inquiry. And this is why somatics become the most important ingredient, and why an embodied critical inquiry is so difficult. But there are those who either already have a very deep relationship with their body or want to develop a very deep relationship with their body, and an embodied critical inquiry turns out to be a viable option. That I pair it with an embodied form of narrative construction is merely to demonstrate how embodied knowledges can play out in a fully epistemological, and truly communicative, sense all the way down.

Somatic training and somatic practices are exercises that operate best as priming functions for an embodied critical inquiry because one needs to become very centered on their kinesthetic being before they can tap into embodied knowledges. Somatic exercises are a way of pulling one's attention away from the external world and directing it inward. This is why somatic skills typically require training and why practices like an embodied critical inquiry will be more or less successful, or require more or less non-kinesthetic prompts and instructions. Generally speaking, one can thinking of somatic work as pulling oneself into deeper and deeper layers of the body. The deeper one aims to go into the body, the more challenging the process becomes. Being so focused as to withdraw from external stimuli, or to develop different kinds of relationships to them—such as the relationship the stimulus makes with the body—is strenuous. Moving deeper towards one's intuition is hard because one has to relax into the body and let the body lead. This is not improbable since many people operate on bodily intuition regularly; what makes this difficult is that one needs to sustain the kinesthetic focus as they connect with their embodied intuition. The most arduous step in somatic work, at least for this purpose, is intentionally tapping into the embodied knowledge. Embodied knowledge is something that can stay on the surface for some, it is something that will reveal itself in important situations, and it can kick in during activities that require habituated procedural knowledge. Yet it can be unbelievable punishing to reach in and down and knocking on the door of one's embodied knowledge to attempt a direct conversation with it using one's sensory modalities. But this is exactly what somatic work primes us to be able to do. This is one of the reasons why it is so effective in therapy.

When I facilitate this process, I have a particular order of operations. I begin with static somatic work, the transition to minimally moving somatic exercises, and then immediately into improvisational work (as will be discussed later). Somatic work links the person from the concept to the embodied knowledge. The best way that I can describe it is that one basically marinates their body in the concept. That is, one aims to take the explicit verbal version of the concept and redirect and rearticulate its meaning through the body. One must labor to take the concept they are holding "in their mind" and move it into the body, where it can resonate beneath the surface and in relation to the various kinesthetic sensations one directs their attention towards. I then guide the mover into some very light movement to lighten the burden of the external stimuli on their kinesthetic sensibility. This is the point at which I abet the mover into a primarily thinking-in-movement mode—they begin to move with the concept as it has become embodied. On my version, non-"dancing" somatic work activates one's sensory capacities and brings affect about the concept closer to the fore of one's embodied awareness. The connection to, and focus on, this particularized affect helps shape faint felt relations that one has to the concept. This is the moment when improvisation comes into play.

Moving from somatic work to improvisational movement is the/a next step in the embodied critical inquiry process. Somatic work is a way of getting in touch with your affective relationship to the concept and improv is a way of using your body to articulate these connections. When a mover is solidly in their kinesthetic-cognitive place, they can allow their movement to be kindled by their embodied knowledge. This is why I term this mode of knowing as a movement-knowing, because the knowing is unveiled through the movement and on to the body. The body is literally a palimpsest for embodied knowledge that becomes inscribed by movement imbued with somatically induced conceptual-affect, or an embodied-conceptual felt sense.

The body writes on the body through embodied metaphors. This is how and why it is possible to "read" the movement, as I have been saying. Daniel Nagrin speaks to the ability of a dancer to communicate directly through movement metaphors:

> In one of our sessions, I asked Maggie to take the floor—alone. When she was there, standing, waiting, I asked her, "Maggie—no words. Why do you dance?"
>
> Her first response was a bemused expression, and then she plunged. There ensued a long, detailed, complex, improvisational solo that never for a moment lost its compelling flow and never for a moment provoked the deadly question, "What does she mean?". Her performance convinced me as never before that when a dancer has a firm grasp on a specific image, reinforced by a strong emotional context with the imagination to find the right movement metaphors, questions of meaning, of intent or content, do not arise in the minds of the viewers. Everyone is engaged and "knows" what they are seeing, constructing a coherent flow of sensations, feelings and conceptions that run parallel and close to what they are witnessing. (Nagrin 1994, 64)

Embodied metaphors can convey the content of the embodied knowledge if the movement is sufficiently somatically grounded. These dancing metaphors are the expressions of movement-knowings. However, they are not symbols, as Langer argued, since they are not virtual. Non-scripted dancing is directly cognitively communicable through embodied metaphors. When one moves in accord with embodied intuition, the body will suggest certain ideas in its shape and dynamics. When one moves in accord with movement-knowings, the body tells you what you think. This is what I have been referring to when I have been purporting that an embodied critical inquiry is a form of working in reverse of narrative choreographing using embodied metaphors. We are able to use embodied metaphors purposefully to communicate narrative to others in a relatively objective manner (Welch 2019). Because embodied metaphors can be studied and learned as a form of vocabulary, one can look at their improvisational movement to identify the surfacing embodied metaphors to identify the content of the embodied knowledge. As I show in the methodology, this step in the process is but a mere couple of worksheets away from identifying the embodied critical questions mined from the body through movement.

Another reason why somatic training is so important is that this kind of training can help one become more fluent in reading embodied metaphors in the critical inquiry processes. When people use dancing metaphors to construct communicative narrative, the best way to be effective and objective is to aim towards the most common denominators, which would be movements and dynamics that are primary metaphors; these are shared more broadly and have fewer contextual or cultural inflections. Because blended conceptual metaphors are culturally informed and so bear cultural content, there's a sense in which embodied metaphors can be a bit like a dialect. Primary metaphors, on the other hand, have less cultural content and are more immediately informed by, and enacted in accord with, sensorimotor contingencies. However, when one is engaged in a somatics-based improvisational method for an embodied critical inquiry, the embodied metaphors follow from the mover's subjective movement-knowings. There is a sense in which almost no one aside from the mover can accurately read the embodied metaphors for their full cognitive content because movement and dynamics are very idiosyncratic and one's relationship to a concept is unique. No two people will dance the same idea in the same way. So it is advantageous for the mover to have some version of a somatic training so that they can become more attuned to their kinesthetic being. Somatic training helps movers learn to better "read" their thinking-in-movement. This is also why it is important to have a trained facilitator guide one through somatic work since learning to understand languages—even your own—can be obstruse on one's own.

Finally, somatic training helps movers identify and break ingrained socially constructed movement habits that impede the body's ability to move intuitively from a place of implicit knowledge. This is important for an embodied critical inquiry in a way that it is not advantageous for narrative construction. For one to be able to properly read their own embodied metaphors, they have to be able to shake the tendency to enact standard and commonly constructed body dynamics; it is imperative that the mover get in touch with their own mindbody's distinct affective articulations. But equally important, one needs to be able to put aside preconceived meanings associated with particular embodied metaphors when attempting to read their movement. One might say that a person's inclination to immediately interpret their movement in keeping with the dominant "narrative" of embodied metaphors is a form of embodied implicit bias. If one has very little connection to their kinesthetic being, they will have more barriers to discerning movement that is meaningful content in the movement assessing process; they will more than likely pick out movement that "looks" important. This is why it can sometime be more difficult to facilitate a mover who moves strictly to entertain because, like many star students in an intro to philosophy class, they may be liable to give answers someone would want to hear to "wow" them.

Embodied Dialogues

An embodied critical inquiry—an Inquiry-in-Movement—is just another class of thinking-in-movement. Just as we all think in movement, we all also inquire in movement. But equally so, just as movers—such as dancers—can intentionally try to cultivate more precise exercises to think in movement, and to see that thinking on the body, we can cultivate practices and skills to purposefully question through movement. One fascinating component of this capacity to think and inquire in movement is our ability to do so with others. While I attempt to save the enactivist framework that explains how this is possible for the next chapter, here I show how somatics-based improv can operate as the basis for an embodied dialogue between movement partners qua participatory sense-making.

Given that I focus this chapter on the epistemology of thinking-in-movement, and given that this section addresses the possibility of embodied dialogues during somatic-based improvisational exercises, it becomes important to address the epistemological body. Up to this point, I have spoken about the becoming, phenomenological body and its embodied knowledge; but this is not the extent of the epistemological body. As I have shown, thinking-in-movement is a moving epistemology. Important to this moving, procedural epistemology is a moving, procedural body.[13] The reason why an explicit discussion of the procedural body becomes pertinent here is that embodied dialogues rest on participatory sense-making, which itself rests on a procedural—enactive—account of social cognition. These frames make it possible to evidence how bodies can, and do, share

[13] Other theorists speak of a processual body. This is a similar account of the body but tends to be more of an ontological claim than an epistemological claim. These accounts of a processual body follow from the metaphysics of Whitehead and Bergson. In my work here, both Manning and Massumi talk about the processual body but they also do so specifically for the political impact of this kind of body. While my account of the procedural body is compatible with the processual body, it should not be conflated with it.

knowledge qua body-body touch. Bodies can communicate beyond gesture, which is typically conceived as the grounds of embodied communication; bodies can communicate at the level of affect, tension, energy, and synergy. Somatic exercises allow for partners to communicate between their bodies at this deeper, cognitive-affective level. And this requires an understanding of the moving, procedural body as epistemological.

One of the ways to see the procedural, epistemological body is by looking at how others have described the ontological body as a force—"[bodies] have more in common with hurricanes than with statues" (Di Paolo et al. 2018, 7). When we look at the body in all of its dynamical interactions and productions, we can not only see its fervent—and chaotic—becoming, we can also see body-force as a kind of moving know-how. This helps clarify claims that the body's knowledge often outruns, or is in excess of, understanding. Erin Manning (2003) captures the force of the body when she speaks of the processual body as a "collision" of processes. But Phillipa Rothfield (2019) gives us a phantastic account of the improvisational body using a Nietzschean analysis of force in the will to power. Her analysis can be read as pointing towards the epistemologically procedural body qua process through her argument that improv is a body's achievement. She says, according to Nietzsche, that force is the beginning of all things, and the body is the origin of impulses that are perpetually in flux. Agency is between the forces themselves and this interaction between competing forces—the will to power—generates movement (94). The value of improv lies in its kinesthetic variation and also in its very nature as an unscripted becoming that emerges from the body's flux of impulses and forces. She explains:

> The Nietzschean perspective looks upon the dance as a site of multiple twists and turns, involving choices embedded within the action itself rather than decisions made by some sovereign agent (the dancer) …If not the work of the dancer but the work of the dance itself, improvisation becomes reconceptualized as a shifting play of force. There is an element of chance within this play—chance lays the groundwork, sets the field of possibility, for that which follows. (90)

Because the body is stirred by these forces to pursue openings through its know-not-how within the improvisational process, the body enacts a process of experimentation that transcends conventional normative practices (99). Rothfield's argument brings us back around to Manning's application of the bodied collision to the epistemology of dancing:

> When movement dances in the dancing, what we experience is a lived extraction from the plane of immanence that is total movement, leaving us in the vibration of what is beyond the predetermined body, in the realm of movement-thinking. And it is here, in the thinking in act of movement-moving, that movement begins to compose (us). (Manning 2003, 15)

And even Sheets-Johnstone:

> The dance comes alive precisely as the dancers are implicitly aware of themselves in the form, such that the form moves through them: they are not agents of the form, but its moving center. Because they are themselves immersed in what they are creating, because they are not going through specified movements as one would go through a series of technical maneuvers, what is created is what appears as a unique interplay of fluid, ever-changing forces, denying dynamic and cohesive flow of energy, not in the sense that the dancers continually change relationships and positions, but because the dancers and the dance are one. (Sheets-Johnstone 2015, 3)

Process in being and knowing are products of the body's ability to dynamically shape and structure the world and our identities. The body's procedural forces, in part, create and investigate movement affordances.

But the procedural nature of bodies, and their body-forces, have to exert their impulses externally if they are going to be able to ground embodied dialogues. That is, moving, procedural bodies must be moving, social bodies to be communicative at the somatic level—we need social bodies to engage in curiousing with others. This claim is sensical when we consider the phenomenological notion of flesh. If at least some part of agency is a result of the collision of forces and the competition between them, such as this impartial understanding of the will to power, it is reasonable to locate such forces as overrunning the skin. When we mesh Manning's recognition of the body's knowledge trickling out of the body with this idea of body-force, one can postulate that the body's agential forces, which contains its knowledge, can merge with the forces of another body across the relation of flesh through touch since touch grounds the procedural body. Because touch is an act of reaching toward, and because touch is one social way of traversing the phenomenological flesh of relationality, touch itself spurs processes of worlding when bodies move. Manning attests that "I reach out to touch you in order to invent a relation that will, in turn, invent me" (Manning 2007, xv). The understanding of bodies as social all the way down to the affective layers helps us bridge somatic work to relationality through participatory sense-making. The succession from the soma to the relational and back again serves as the conduit for somatic-based embodied dialogues.

Enactivists refer to this form of a procedural epistemological social body as a linguistic body (Di Paolo and De Jaegher 2017; Di Paolo et al. 2018). A linguistic body undergirds the phenomenological body's ability to engage in sense-making through gesture, embodied metaphors, and shared understandings; this is a mode of social cognition hermeneutics (Gallagher and Allen 2018). When we take the body as a linguistic agent, social interaction can then be defined as the regulated coupling of the relational dynamics between participants. When we think of social interaction as a project of bodied coordination spurred by gestures and utterances—which can themselves be movements—we can think of social interaction as an embodied, ongoing process (Di Paolo and De Jaegher 2017, 492–493). Social action via the linguistic body is largely regulated through social modes of cognition, which rely on intersections between sensorimotor activities, affect, and affordances; social interaction relies on perception, which is itself social. The key mechanism through which we regulate our couplings with others is through the uptake and interpretation of affordances made available to us by others' emotions and actions (Gallagher and Allen 2018, 2614). One of the examples Gallagher and Allen supply that helps communicate the embodied depth from which we perceive and regulate socially is that of arousal. They explain that changes in one's arousal is constitutive of perception and these changes in arousal (affective, metabolic) have an effect on both the perceiver's and the perceived's body (ibid.). Because of the role of affordances in perception, our coordination with others is anticipatory rather than reactionary (2642). Together, these ideas reveal how somatic work can stimulate communicative avenues through body-body touch. Embodied dialogues, which can be construed as a shared, practical, embodied understanding, are a function of body-body attunement within the "interaffective space" (Tewes et al. 2017, 5). Manning would refer to the dynamic sustaining embodied dialogue as "a tuning not of content but of expression-with" (Manning 2007, 11).

Because bodies anticipate, and respond to, one another from the dynamics of the body vis-à-vis arousal, bodies can perceive other's bodily communications down to, and through, the soma; I argue that this is largely possible through the syncing of a partner's perceivable interoception sensations. Fraleigh (1987) emphasizes that dance closes the distance between the self and other; somatic work would help to functionally close this distance. As if in response to her, Albright insists that we must let go of our ingrained territorial approach to bodily integrity. In order to become properly and fully responsive, she argues that we must recognize ourselves as conduits for, and to, others (Albright 2013, 239). Most important to understanding how bodies can attune through somatic body-body touch is her claim that contact creates an empathetic kinesthetic perception that bridges our subjectivities (242).

When somatic work closes the bridge between self and other, the individuals are able to conjunctively cultivate what Braude terms "empathy in the flesh" (Braude and Shulman 2015, 130). From a phenomenological perspective, the best way to describe this bridging of subjectivities via each partner's empathy in the flesh is to hark back to Merleau-Ponty's chiasm. Because partners can attune to one another through body-body touch, they close the gap of the flesh and therefore enter into the space of the chiasm. This, in short, means that just as the chiasm reflects our experience of the touching-touched hand, somatic attunement can lead to a merging of embodied dynamics that result in a very similar sort of touching-touched shared subjectivity constituted by the interaffective space. Midgelow argues that partner improv occupies the chiasm through reciprocal perceiving/perceiving that (Midgelow 2018, 124). My claim is that somatic work can bring that reciprocity deeper into, and between, the bodies.

In addition to somatic work, cognitive theorists also explore the extent to which dance operates as a mode of participatory sense-making. April Flakne surveys contact improv as a practice of participatory sense-making. She argues:

> By destabilizing proprioception, together with marked, habitual, and socially mediated gestures, contact improvisation has the capacity to tap into the intercorporeal origins of even instrumental and communicative gesture, offering special access to the presubjective, affective, and social processes that constitute the deepest layers of social ontology. (Flakne 2019, 527)

On her account, contact improv opens up affective and kinesthetic agency through the movers' sense of alloception. Flakne defines alloception as an enactive bodily receptivity to sensory and kinesthetic processes of others akin to mind reading/other's minds. She identifies it as an active form of kinesthetic empathy that vitalizes our potentials and capacities for respondability to our partners; it is an enactive responsiveness to another consequent of creative interactive gesturing (535). And Michael Kimmel (2019) explores how tango, specifically, simultaneously creates and sustains micro skills of coordination. He introduces the process of bodied coordination as participatory sense-making using tango to reveal the scope of creative sense-making through highly structured movement activities. In tango, the creativity resides in the accentuation and ornamentation—what some of us call the "flare"—the autonomous individuals add to the coordinated movement patterns (564). Because even highly structured dances offer affordances, the idiosyncratic and creative flares in the moment present solicitations to the partner for innovative choices to maintain the pairing and move it forward.

Buttingsrud rightly attests that contact improv is a process of two-fold bodily-affective decision-making (Buttingsrud 2021, 11). If we look back to the social cognition model of the linguistic body, we can see the embodied cognition basis of embodied dialogues qua improvisational movement being the coregulation between the bodies:

> Coregulation implies accepting a more direct participation by others in our own sense-making without entirely giving up our agency...we exert some influence on something we *can* control ourselves: our willingness to accept the participation of the other in constructing a joint social act...coregulation is not only acting together but giving the interactive situation itself influence on our acts (regardless of whether we do this consciously or not). All of our acting requires a certain degree of letting things be ... [but] a social encounter that is constantly changing gears is hard to sustain and can exact ahigh emotional cost in frustration, alarm, and confusion. (Di Paolo et al. 2018, 149, 152)

However, as I've intimated previously, shared sensorimotor sense-making vacillates between bodies; that sense-making is reciprocal entails that there are going to be alternating asymmetries in attunement contributions.

But just as with any social interaction, embodied dialogues can involve asymmetries that are not neutral and equally weighted. That is, as in any procedural epistemological interaction that involves sharing knowledge, bodies can enact and imposes biases on one another; and as happens in typical social interaction, bias imposition can threaten a partner's autonomy and coregulation can break down until one is dominant and no longer gives uptake—is attuned to, and by—to the other. Microaggressions and practices of domination exist just as strongly within movement as they do within verbal language and social practice. When relations of oppression obtain between partners, even when unrecognized, implicit biases can materialize through, and in, movements such as posture, gaze, and leading. In cases such as these, embodied dialogues "breakdown".

Although embodied dialogues require a blending of embodied energies, as it were, coregulation and participatory sense-making requires that individuals retain their discrete subjectivity and autonomy. One exceedingly common example of the possibility for autonomy, and thus dialogue, to break down in dancing is the overzealousness of the leader, who cuts off any kind of invitation for the partner's creative participation as a an interlocuter (Kimmel 2019, 571). If the partners somehow melded into one weird synergistic blob, there would be no "dialogue". While I'm not entirely sure that somatic work can literally enmesh dancer's somatic layers, there is some sense in which we can imagine that partners overdo it; this might suggest something like an over investment of the other, such that the other does not intend to become the regulator, but the mover seems to give themselves over to their partner. So there are a number of ways in which partners enact coordinated patterns but fail to sync properly: too much or forced synchronization and dependence/subordination all lead to the diminishment of each partner's autonomy and impedes the communicative capacities of somatics and improvisation. Galbusera et al. point to the dangers of sustaining your own intrapersonal synchrony in tandem with the interpersonal synchrony involved in partner work, such as dance (Galbusera et al. 2019, 14697); on their account, the higher the interpersonal synchrony, the more diminished the positive affect of the interaction. Shared movement dynamics significantly alter one's own affect and so over-doing it in some way can lead to an inability to regulate on one's own, which undermines the point of a participatory sense-making activity.

Bibliography

Albright, Ann Cooper. "Feeling In and Out: Contact Improvisation and the Politics of Empathy". In *Engaging Bodies: The Politics and Poetics of Corporeality*. Middletown: Wesleyan University Press, 2013: 237–245.

Barbour, Karen. *Dancing Across the Page: Narrative and Embodied Ways of Knowing*. Chicago: Intellect: University of Chicago Press, 2011.

———. "Dancing Epistemology: Situating Feminist Analysis". In *Back to the Dance Itself: Phenomenologies of the Body in Performance*. Sondra Fraleigh (ed.). Urbana: University of Illinois Press, 2018: 233–246.

Braude, Hillel and Ami Shulman. "Radical Somatics". In *Moving Consciously: Somatic Transformations through Dance, Yoga, and Touch*. Fraleigh, Sondra (ed). Urbana: University of Illinois Press, 2015: 124–134.

Buttingsrud, Camille. "Bodies in Skilled Performance: How Dancers Reflect Through the Living Body". *Synthese* 199 (2021): 1–20.

Di Paolo, Ezequiel, and Hanne De Jaegher. "Neither Individualistic nor Interactionist". In *Embodiment, Enaction, and Culture: Investigating the Constitution of the Shared World*. Christopher Durt, Thomas Fuchs, and Christian Tewes (eds.). Cambridge: MIT Press, 2017: 87–105.

Di Paolo, Ezequiel, Elena Clare Cuffari, and Hanne De Jaegher. *Linguistic Bodies: The Continuity between Life and Language*. Cambridge: MIT Press, 2018.

Flakne, April. "Contact Improvisation and Embodied Social Cognition". In *Oxford Handbook of Improvisation in Dance*. Vida Midgelow (ed.). New York: Oxford University Press, 2019: 527–544.

Foster, Susan Leigh (ed.). *Corporealities: Dancing Knowledge, Culture and Power*. New York: Routledge Press, 1996.

Fraleigh, Sondra Horton. *Dance and the Lived Body: A Descriptive Aesthetics*. Pittsburg: University of Pittsburg Press, 1987.

Galbusera, Laura, Michael Finn, Wolfgang Tschacher, and Miriam Kyselo. "Interpersonal Synchrony Feels Good but Impedes Self-Regulation of Affect". *Scientific Reports* 9 (2019): 14691–14702.

Gallagher, Shaun, and Micah Allen. "Active Inference, Enactivism, and the Hermeneutics of Social Cognition". *Synthese* 195 (2018): 2627–2648.

Grove, Robin. "Show Me What You Just Did". In *Thinking in Four Dimensions: Creativity and Cognition in Contemporary Dance*. Robin Grove, Catherin Stevens, and Shirley McKechnie (eds.). Melbourne: Melbourne University Press, 2005: 37–49.

Grove, Robin, and Shirley McKechnie. "Introduction". In *Thinking in Four Dimensions: Creativity and Cognition in Contemporary Dance*. Robin Grove, Catherin Stevens, and Shirley McKechnie (eds.). Melbourne: Melbourne University Press, 2005: 18.

Hanna, Judith Lynne. *Dancing to Learn: The Brain's Cognition, Emotion, and Movement*. London: Rowman and Littlefield, 2015.

Hawley, Katherine. "Success and Knowledge-How". *American Philosophical Quarterly* 40, no. 1 (2003): 19–31.

Johnson, Mark. *In the Body, in the Mind: The Bodily Basis of Meaning, Imagination, and Reason*. Chicago: University of Chicago Press, 1987.

———. *The Meaning of the Body: Aesthetics of Human Understanding*. University of Chicago Press, 2007.

Katan-Schmidt, Einav. *Embodied Philosophy in Dance: Gaga and Ohad Naharin's Movement Research*. London: Palgrave Macmillan, 2016.

———. "Dancing Metaphors; Creative Thinking Within Bodily Movements". *Proceedings of the European Society for Aesthetics* 9 (2017): 275–290.

Kim, Na-ye. "Identifying Choreographic Knowledge: Choreographic Techne and Phronesis". *Research in Dance Education* 17, no. 1 (2016): 42–59.

Kimmel, Michael. "A Cognitive Theory of Joint Improvisation: The Case of Tango Argentino". In *Oxford Handbook of Improvisation in Dance*. Vida Midgelow (ed.). New York: Oxford University Press, 2019: 563–593.

Manning, Erin. *Always More Than One: Individuation's Dance*. Durham: Duke University Press, 2003.

———. *Politics of Touch: Sense, Movement, Sovereignty*. Minneapolis: University of Minnesota Press, 2007.

Midgelow, Vida. "Improvisation as Paradigm for Phenomenologies". In *Back to the Dance Itself: Phenomenologies of the Body in Performance*. Sondra Fraleigh (ed.). Urbana: University of Illinois Press, 2018: 59–77.

Mojica, Monique. "Stories from the Body: Blood Memory and Organic Texts". In *Native American Performance and Representation*. Tucson: University of Arizona Press, 2009, 97–122.

Nagrin, Daniel. *Dance and the Specific Image: Improvisation*. Pittsburgh: University of Pittsburgh Press, 1994.

Pakes, Anna. "Knowing Through Dance-Making: Choreography, Practical Knowledge, and Practice-as-Research". In *Contemporary Choreography: A Critical Reader*. Jo Butterworth and Liesbeth Wildschut (eds.). New York: Routledge Press, 2009: 10–22.

Parviainen, Jaana. *Bodies Moved and Moving: A Phenomenological Analysis of the Dancing Subject and the Cognitive and Ethical Values of Dance Art*. Tampere University Press, 1998.

Pezzulo, Giovanni. "The Contribution of Pragmatic Skills to Cognition and Its Development: Common Perspectives and Disagreements". In *The Pragmatic Turn: Toward Action Oriented Views in Cognitive Science*. Andreas Engel, Karl Friston, and Danica Kragic (eds.). Boston: MIT and the Frankfurt Institute for Advanced Studies, 2015: 19–33.

Rothfield, Philipa. "Chance Encounters, Nietzschean Philosophy, and the Question of Improvisation". In *Oxford Handbook of Improvisation in Dance*. Vida Midgelow (ed.). New York: Oxford University Press, 2019: 89–102.

Rucińska, Zuzanna, and Kenneth Aggerholm. "Embodied and Enactive Creativity in Sports". In *Handbook of Embodied Cognition and Sport Psychology*. Massimiliano Cappuccio (ed.). Cambridge: MIT Press, 2019: 667–694.

Schaeffer, Catherine. "Living Shin". In *Moving Consciously: Somatic Transformations Through Dance, Yoga, and Touch*. Sondra Fraleigh (ed.). Urbana: University of Illinois Press, 2015: 75–92.

Sheets-Johnstone, Maxine. *The Primacy of Movement.* Philadelphia: John Benjamins Publishing Co., 1999.

———. *The Phenomenology of Dance.* Philadelphia: Temple University Press, 2015.

Stevens, Catherine, Stephen Mallock, Shirley McKechnie, and Nicole Steven. "Choreographic Cognition: The Time-Course and Phenomenology of Creating a Dance". *Pragmatics and Cognition* 11, no. 2 (2003): 297–326.

Sutton, John. "Moving and Thinking Together in Dance". In *Thinking in Four Dimensions: Creativity and Cognition in Contemporary Dance.* Robin Grove, Catherin Stevens, and Shirley McKechnie (eds.). Melbourne: Melbourne University Press, 2005: 50–56.

Tewes, Christian, Christopher Durt, and Thomas Fuchs. "Introduction: The Interplay of Embodiment, Enaction, and Culture". In *Embodiment, Enaction, and Culture: Investigating the Constitution of the Shared World.* Christopher Durt, Thomas Fuchs, and Christian Tewes (eds.). Cambridge: MIT Press, 2017: 1–21.

Todres, Les. *Embodied Enquiry: Phenomenological Touchstones for Research, Psychotherapy, and Spirituality.* New York: Palgrave Macmillan, 2007.

Van Manen, Max. *Phenomenology of Practice: Meaning Giving Methods in Phenomenological Research and Writing.* New York: Routledge Press, 2014.

Wallis, Charles. "Consciousness, Context, and Know-How". *Synthese* 160 (2008): 123–153.

Welch, Shay. *The Phenomenology of a Performative Knowledge System: Dancing with Native American Epistemology.* New York: Palgrave Macmillan 2019.

Wengrower, Hilda. "The Creative-Artistic Process in Dance Movement Therapy". In *The Art and Science of Dance/Movement Therapy: Life is Dance.* Sharon Chaiklin and Hilda Wengrower (eds.). New York: Routledge Press, 2016: 13–32.

CHAPTER 6

Enactivism and Embodied Critical Inquiry

Because most of Western philosophy devalues the body and dismisses the idea of embodied knowledge, many dancers might believe that embodied knowledge, because it is so intuitive, is in some sense mysterious. But as I have been at pains to show throughout the book, this is an unfortunate result of epistemological bias and hermeneutical gaps created by this sequestering of the body to the disposable and intrusive. As all of the embodied cognition and phenomenological theorists I engage thus far have attested to, there is truly nothing random or mysterious, or truly hidden from us, with respect to our embodied knowledge. From affect theory to body logics to decision-making in action, one thing that should be apparent at this point is that our bodies know, and they know because they think, and they think because they move. It is for this reason specifically that we are able to know what our bodies, and what other people's bodies, are saying and choosing when they move in specific sorts of ways. Throughout I have been engaging these theorists in a sort of piecemeal fashion to attempt to parse and sort the various ingredients needed for an embodied critical inquiry. In this chapter, I make a pointed effort to draw all of the insights together to make a specific sort of claim: while most embodied theorists are amenable to the idea of dance as thinking and even dance as inquiry, thinking-in-movement, however one enacts it, is specifically enactivist.

That is, while most, if not all, embodied cognition theories can help us understand these ideas which seem so evident to those of us who move,

S. Welch, *Choreography as Embodied Critical Inquiry*,
https://doi.org/10.1007/978-3-030-93495-8_6

but largely remain elusive, I believe it will be the enactivist model that can best account for how and why thinking-in-movement is so and thus, right and accurate. And I have intimated throughout the previous chapters, the various kinds of embodied cognition that have weaker and stronger commitments to embodiment are not necessarily so sharply differentiated. It is true that it is important for scholars to establish boundaries for theoretical purposes between camps, but interaction with live movement reveals that sometimes, in some places, these boundaries become more permeable in ways that do not undercut the embodied commitments of those specific theorists. My approach to embodied cognition is a strong enactivist one; but I would not be able to engage dance from an enactivist perspective without embodied metaphors. I think that what has demarcated these two from one another has been the focus on verbal language as the site of analysis. I certainly think that brain maps and mirror neurons are essential to making sense of embodied sense-making. But I do think that one's understanding of some of the aspects of these schemas will somewhat evolve and amend if they look for metaphors on the body rather than looking for the body in our metaphors. It seems to me that when one moves from verbal language to true body language qua moving body that some of the barriers between functionalist and enactivist versions relax.

So there is a two directional benefit from both this project generally and this chapter specifically. An articulation of an embodied critical inquiry from an enactivist perspective can help movers understand explicitly and in detail what they already presume to be true; *and* demonstrating how thinking-in-movement works in application through an easily repeatable method that can be examined and explored gives embodied cognition theorists, especially enactivists, more reason to take dance more seriously. Additionally, movement methods such as the one I include are more accessible for empirical, phenomenological analysis. Most research on dance and embodied cognition include the use of fMRIs, which makes the examination of dance almost wholly inaccessible as a research mode since fMRIs only capture the brain's responses to a person's viewing or imagining or remembering of dance. It is not yet possible to conduct an fMRI on a moving, dancing subject. But fMRIs are not the only way to produce reliable outputs. If empirical embodied cognition scholars were to use living movers in conjunction with robot research, they may get a broader, brighter picture of how the neurological maps onto the phenomenological for a richer neurophenomenological account.

In this chapter, I demonstrate how the phenomenology of dance counts as a naturalized phenomenology, such as enactivism. Enactivism falls under the domain of a naturalized phenomenology given its melded lineage of Thompsom, Varela, Dewy, and Merleau-Ponty. Additionally, to expand on the section regarding embodied dialogues in the previous chapter, I provide a thorough summation of Di Paolo et al.'s theory of linguistic bodies, since I find this enactivist model to be best suited to understand how bodies can communicate all the way down. A rendering of bodies as fully communicative is indispensable for understanding how somatic exercises can operate as conduits to embodied knowledge, which can then be expressed onto the body through movement. Most importantly, I revisit embodied metaphors. I switch to the Katan-Schmidt's version of dancing metaphors to corroborate my contention that embodied metaphors are categorically compatible with enactivism. If my method turns on the readability of the body, and thus literal embodied metaphors, I must substantiate that these two elements are congruent. Without this bit, my purpose of providing a philosophical formulation of an embodied critical inquiry virtually breaks down.

Dancing as Naturalized Phenomenology

The goal of naturalism is to transfer our understanding of the mind to the natural sciences. For example, theorists might aim to describe the activities of the mind as mental states that can be identified with neurological states. But phenomenologists have decried such attempts as missing the point of the mind's role in examining the world; and that attempts made to naturalize consciousness erase the experience of consciousness. But some phenomenologists are now moving in a direction that merges the ostensibly disparate approaches. According to Gallagher and Zahavi (2012), the idea of a naturalized phenomenology stems from Merleau-Ponty's discontent with the stark division drawn between transcendental philosophy and empirical science. He believed that much empirical work could be enhanced by analyzing the explanandum through a phenomenological lens. His interest was less in whether phenomena could be naturalized reductively and more in whether the phenomenological method itself could be naturalized (32). Phenomenology and cognitive science can be used in tandem to shed light on the nature and reality of experience. For this reason, many embodied cognition scholars follow the work of Francisco Varela et al., who argue that we ought to suspend our beliefs or theories about experience and, instead, work to gain more immediate insight into experience to yield better descriptions of it. A naturalized phenomenological method would design experiments that incorporate, and/or are framed by, subjective input going in and coming out of the testing (44).

Central to a naturalized phenomenological method, and to this account of an embodied critical inquiry, is the phenomenological explanation of the cognitive experience of reflection. Here, reflection amounts to one's ability to turn inward to make their experience the object of consideration. Because reflection can disclose meaningful knowledge about one's experience to a knower, reflection can always transform experience. Gallagher and Zahavi explain that the relationship between pre-reflective consciousness and reflective self-consciousness consist of a dialogical, interpretive process; when we reflect on our experience, it already involves our pre-reflective consciousness, which can substantiate our interpretation of it (71). Consequently, reflection turns out to be a precondition for self-critical deliberation (72). This construal of the relationship between pre-reflection and reflective self-awareness is necessary for an embodied critical inquiry since the mover must be able to engage their pre-reflective movement-knowings with a strong reflective self-awareness. But reflection in the moment is also important given its role in deliberation, which dancers engage in during the dancing itself. Reflection, then, anchors the embodied hermeneutics constituting the phenomenology of practice and Practice as Research, which frames the moving epistemology of thinking-in-movement. Dance is an ideal archetype of enactivist naturalized phenomenology because of the role of pre-reflection, reflection, and self-reflection in the doing of the activity.

Sheets-Johnstone, as Varela and others do, speaks about the doings of dance as a kind of transcendental reduction because she argues that we must suspend our beliefs and judgments about dance to actually dance. In dance, we are "too alive", "too vibrant" (Sheets-Johnstone 2015, 1). Relatedly, enactivism positions the core of meaning-making within the body-brain-environment dynamical system. As such, perception is something that we are always actively doing, even when it seems passive, because we are always framing, engaging, and responding to our surroundings—specifically as they have to do with our current body orientation and objectives. Sheets-Johnstone captures this idea as it is within dancing because dance requires that we are purely engrossed in, and responsive to, what is happening to us, around us, and because of us. She rightly attests that the meaning of dance only comes alive if one is having the lived experience of the dance (2). Dancers create and direct the meaning of dancing just by virtue of their being intent upon its "*emerging* form" and "engrossed in its *unfolding*" (1, italics mine).

Yet, in what seems to be a surprising move, Sheets-Johnstone explicitly claims that "reflection" impedes meaning construction and enactment in dancing. She claims that the doing of dancing is, and should be, free from reflection lest it overwhelm the intuitive experience of creative movement. Fraleigh makes a similar move to this; she says:

> The body-subject can be sensed in dance and through the dancer when she is unified in action; that is, when she is not reflecting on herself or her action but living the present-centered moment in her dance as a unity of self and body in action. (Fraleigh 1987, 13)

On Sheets-Johnstone's account, practices of reflection in dancing generate distance between oneself and the dance. She claims that reflection on dance spurs a "lifeless fact"—a kind of second-hand account—of its significance (2). As she sees it, the role for reflection is not in dance, it is about dance. That is, for Sheets-Johnstone, reflection should be engaged afterwards to analyze and better understand the meaning of the dance (4).

However, I argue that a problem inheres the notion of reflection that stems from cognitivist conceptions that preclude the active and deliberative role of the body itself as a mechanism for reflection. This more traditional version of reflection is backward looking and not live or online. Expert skill execution recognizes one's ability to make their current experience an object of reflection for deliberation during movement. Fortunately, Fraleigh also moves in this direction. She claims that it is important to recognize the body-subject as a body-object, which requires that one take an objective attitude towards the moving body whilst being careful not to objectify it. Further, Fraleigh is explicit about the importance of pre-reflective stance during dancing as a way to listen to the body holistically (Fraleigh 1987, 15). This pre-reflection allows the dancer's consciousness to select and sort through purposeful perception; I would designate this kind of perception as a pre-reflective cum reflective perception rather than solely pre-reflective perception. When we reconceive of pre-reflection, reflection, and self-reflection as phenomenological reflections associated with Noë's action in perception, Rucińska and Aggerholm's enactive creative responsiveness, and Montero's thought in action, the role for reflection in dancing skirts the difficulties with reflective practices that concern Sheets-Johnstone and Fraleigh. Here, reflection is regarded as an activity within action that contributes to the success of, and deliberation within, an action and action-based perception; it is in the moment. This can help one see how the body is simultaneously moving from a place of knowledge and reflecting on that knowledge deliberately in action. Traditional reflection is backwards looking and operates as a kind of objectification of experience. Pre-reflection, reflection, and self-reflection all inhere in live action to determine what comes next. Pre-reflection is the lived phenomenological kinesthetic aspect, reflection is the simultaneous perceptive assessment and evaluation of the options within the horizon, and self-reflection is the component that leads the subject to her choice and adds that real time creativity into the movement choice that is itself imperceptible and barely measurable given its lightning speed time scale. Additionally, one can interpolate the concept of pre-reflection into a the more intelligible notion of implicit knowledge.

For someone embedded in embodied cognition discourses, these accounts of dancing might be identified as an embodied functionalist account. On this account, the body is regarded as a primary vehicle for cognition and sense-making but does so in conjunction with representations; the body is considered either a helper of, or the partner to, the mind as it is conventionally conceived. Functionalist accounts are often regarded as minimally representationalist, which include the action-oriented representations referenced in the introduction.[1] As such, internal states both describe the world and give us affordances (Stewart et al. 2010, 224). This assessment is fair, particularly as it relates to the prevalence of the notion of the virtual in most of the dance phenomenology analyses. On the one hand, many dance phenomenologists speak of the body as a sensorimotor sense-making *instrument*. Sheets-Johnstone deploys one of the core embodied cognition concepts of the body schema (which she also critiques in other places), which she claims is the foundation of our pre-reflective knowledge regarding body spatiality. She rightly claims that our body schemas allow us to grasp gestures and movements meaningfully, which she describes as "a continuous and unified being hereness" and "exists for us as a totality" (Sheets-Johnstone 1984, 17). Relatedly, she

[1] See: Clark (1997). Clark straightforwardly dismisses enactivism as plausible. However, there have been many advancements since then. And enactivists seem somewhat comfortable positing the plausibility of AOR.

takes dance to be a kinetic phenomenon that appears in its aliveness and then takes on self-consciousness (22). Contrary to Fraleigh, she does not think that the body can be taken as an object of reflection during dancing because consciousness is the structure, not the content, of the body's being. But this is not particularly worrisome since it would appear that Sheets-Johnstone's idea of taking the body as an object is more akin to Fraleigh's concern of objectifying the body. And while she purports that the dancer is not aware of herself as she creates the form, she does actually operate consistently with the online account of deliberate expert skill execution insofar as she identities the dancer with the dance (29).

Where these theories seem to backslide towards cognitivism is, I think, in their adherence to Langer's conception of the virtual, which indicates the body to be some kind of generator of symbols and representations through its participation in art. Sheets-Johnstone signifies the form-in-the-making as the fundamental description of the consciousness-body relationship in dance (28); dance, as art, creates and presents an illusion of force—virtual force, which is the form-in-the-making that temporalizes and spatializes the dance. It is a virtual form because the movement is not the authentic movement of life; it is the replication of lived movement plus some adornment. The abstract-virtual is mimesis of the affective-actual. She argues that "if Langer is right that the dancer is a symbol inside the dance that is also a symbol then the dancer as actual consciousness-body must be logically related to a symbol" (35). The dance, versus being in the dance, is a representation. She goes on to explain:

> It is the ecstatic structure of consciousness-body that is the point of congruence between dancer as symbol and as actual consciousness-body. The dance is a symbol of the form of human feelings and is presented by a symbol of the human who fuels those feelings in everyday life. The human who feels his feelings as lived experiences in everyday life is collectively aware of himself through his effective consciousness-body... In dance, consciousness exists its body and movement as a form-in-the-making which is a sheer form, and a spatial unity and temporal continuity exist because at this pre-reflective level, consciousness-body is likewise implicitly aware of its static structure. (37)

More specifically, the image of the form of the affective movement is not an indirectly sensible phenomenon; it is an "imaginative visual-kinetic representation" (95). Where the embodied cognition vis-à-vis phenomenological, sensorimotor meaning-making converges with cognitivism as a whole is in her claim that the dancer, as a symbol, just is the center of force. (38)

Throughout the book, I purport that the use of cognitivist language in dance phenomenology stems from an absence of otherwise-language. But this hasn't prevented cognitive theorists from seeing the enactivist nature of dance phenomenology. And given the functionalist leanings of Sheets-Johnstone's and Langer's language,[2] it makes a whole heaping amount of sense to find their work extensively cited on that side of the embodied cognition scale. However, as more and more embodied cognition theorists and, even, radical enactivist theorists emerge, the use of dance appears to be a significant and highly prevalent context within which such non-representationalist scholars orient themselves. For example, David Best rightly argues:[3]

> We have been for so long and so deeply immersed in the conviction that meaning or significance must somehow be behind the movement or linguistic expression that it is hard to grasp that, on the contrary, the meaning is on the surface…it makes no sense to postulate a ghostly mental process, of which the linguistic expression is a symbol. Similarly, the meaning or significance is the movement, in a particular context. Our actions, and gestures, are not expression of supposed "inner" mental states: on the contrary, such behavior is one's meaning or feeling. (Best 1999, 113)

While many might affirm that dance is a paradigm of embodied cognition, and many would agree that it is enactive-adjacent, some might still be hesitant to concede that dancing is an exemplar of enactivism. Some might point to my above description of the wavering language between embodied and representationalist language in mainstream dance phenomenology.[4]

If, for the time being, one partitions meaning-making through dancing from meaning conveyed through dance, one can see in Sheets-Johnstone's description of the metaphysics of dancing an enactivist model qua kinetic dynamical system. She states:

[2] My introduction to Sheets-Johnstone came from Mark Johnson's work.

[3] See also: Nahrstedt et al. (2007).

[4] See also: Albright (1997) and Marin (1983).

> A dance, as it is formed and performed, is experienced by the dancers of perpetually moving form, a unity of succession, whose moments cannot be measured: it's pasts being created, it's present is being created, it's future awaits creation. Yet, it is not an externally related series of pasts, presents, futures—befores, nows, and afters; it is truly ecstatic, it is in flight, it is in the process of becoming the dance which it is, yet is never the dance at any moment. The dance at any moment is diasporadic, a perpetually moving form whose "moments" are all of a piece. (Sheets-Johnstone 2015, 16)

And in response to the cognitive linguistics, functionalist approach taken by Johnson and others, she seems to find her way towards the otherwise-language that was not extant in her earlier dance phenomenology works:

> Clearly, animate forms, forms which, moving intelligently and in responsively intelligent ways in the world about them, are primed for meaning and those meanings are rooted in archetypal corporeal-kinetic forms and relations...because the starting point for descriptive analyses of embodied image schemata is verbal language, bodily experience enters only after linguistic fact, and then primarily only as it exemplifies linguistic metaphoric practices. While language is shown to refer back to the body in intricate and meaningful ways, the body is not recognized as *being already conceptually alive*, indeed, as being already attuned in fundamental and essential ways to the world and to the nature of its own corporeal-kinetic realities. (Sheets-Johnstone 2009, 220–221, 224, italics mine)

Both claims point towards an enactivist principle of emergence because she exposes how the "non-linguistic", "corporeal concepts" emerge from the body directly. In fact, in The Corporeal Turn, she begins to sound akin to a radical enactivist. In parallel, Colleen Dunagan et al. (2019) develop an account of contact improvisation as an emergent cognitive activity. Emergency theory provides an understanding of systems that explains why dynamical systems cannot be reductive. According to them, improvisation—which is the core of an embodied critical inquiry, especially when done with an ensemble, engenders a system through which a dancer's unified behavior emerges from the confluence of perception, habit, expert skill decision-making, and higher cognitive functions (545). And McKechnie (2005) renders improvisation as a dynamical system insofar as such systems illuminate how change and creativity are the "real laws of nature" (McKechnie 2005, 86).

One concern readers may have with how I link the phenomenology of dancing with enactivism to theoretically situate my account of an embodied critical inquiry is my perpetual sequestering of dancing and dance, as if the two share no inherent relationship in terms of meaning-making. Whether or not the meaning of dance as taken up by an audience or as a choreographed mainstay like Swan Lake is enactivist or not falls outside of the scope of my project. There is a sense in which the division I'm drawing can be made along the same line as "online" and "offline" cognition that many embodied theorists pull on. There are some debates within enactivism, though, as to whether there is any cognition that is actually offline. For this reason, I want to incorporate Gallagher's and Rucińska's (2021) account of mental rehearsal as a possible alternative to the representational slant within dance phenomenology. In order to skirt the dangers of mental representations associated with offline cognition, they formulate mental rehearsal fully as a function of motor imagery. Mental rehearsal can include all forms of imagination—imagining—and so has a direct bearing on cognitive processes of creativity and novelty, especially those that constitute improvisation and processes of choreographing. For them, imagination is not offline since that cognitive process is a highly embodied dynamic process. And this notion of mental rehearsal seems suitable to the representationalist interpretations of pre- and post-performance choreography since both the dancer and the audience are engaged in imaginative practices before, during, and after performance—these are the stages at which representationalist currents in dance phenomenology emerge.

Their analysis of mental rehearsal turns on the distinction they draw between mental representation and mental image. A mental representation is a thing, something possessed in the mind, an entity that is equally product and content. A mental image, they argue, is a motoric process of imagining (10). Imagining is a dynamic embodied processes because it involves perception, movement, the temporality of affordances, affect, and knowledge; it is a cross modal motor process because all of this sensory information emerges from the body-brain-environment system in movement. Even movement itself is an imagining:

> Kinesthesia, as the "internal" side of gestural movement can be generated by perception and imagination, as well as literal movement…[m]y neurons, and my body, go into certain states when I engage in such processes, and these are material processes that we can explain and sometimes measure. (10)

As per usual, the authors explicate the embodied process of imagining through the affordances afforded by pretend play, which is imagination in action—a "prospective performance" where the body enacts a performance that is necessarily tied to a context (14). In short, they maintain that an enactive account of imagination comes out of an embodied know-how that activates via motor images and mental imagery; this involves both neural and body processes that are also both kinesthetic and affective processes (13). It might be possible to transmute the imagining with the virtual in order to yield an enactive account of the meaning of dance outside of the moment of creative enaction. If one conceives of imagination and imagining as a motor image via higher order embodied cognitive process, the classification of embodied creativity and curiousing for developing and accessing movement-knowings as higher order cognitive functions makes more sense. Accordingly, it would appear, then, that if one were to naturalize the phenomenology of dance, one would need the enactivist framework. Or rather, enactivism seems to be the key to naturalizing the phenomenology of dance.

Movement-Knowings and Linguistic Bodies

In this section, I set out Di Paolo et al.'s accounts of sensorimotor contingencies and sensorimotor agency (2017) and linguistic bodies (2018) in more detail. Sensorimotor contingencies and sensorimotor agency are pivotal facets of the enactivist approach. While I have referenced them throughout, I want to draw an explicit connection between sensorimotor sense-making and dancing. In doing so, I highlight another respect in which an embodied critical inquiry is best substantiated within an enactivist schema. In this section, I explicate how sensorimotor sense-making undergirds and generates movement-knowings that are drawn upon during the embodied critical inquiry process. Di Paolo et al. (2017, 181) state that enaction is the transition between microworlds, the time between moments. This idea is specifically relevant to an embodied critical inquiry since the meaning of movement is found across the dynamic, largely in the sweeps and swoops of transition, rather than in any one static frame of a move. Enactivism captures the temporality of the spatiality of the meaning-making of dancing. Further, Di Paolo et al.'s analysis of linguistic bodies naturalizes many of the claims made in the phenomenology of dance, as well as in my own arguments. I purport that it is because we are linguistic bodies that we are able to be fully communicative and readable through movement; I develop an analysis of movement-knowings that links the soma to the linguistic bodies model. Additionally, it is because we are linguistic bodies that we are able to be communicative in, and during, embodied dialogues.

Most basically, cognition emerges from sensorimotor life, which is a life lived in interaction with the world and others. Consequently, we enact the world through perceptually guided actions and the cognitive processes that emerge from movement repetition. Our cognitive system is constituted, then, by the ways in which we are able to be sensitive, and adapt, to motor variations that bring about changes in our sensations and/in the environment. This is the body-brain-environment system at the center of enactivism—cognition is an emergent property of the body in action. Following from, and moving beyond, Noë, the enactivist sensorimotor approach to perception eschews internal representations; instead, perception is a result of sensory and motor patterns that give us access to objects in the world. We perceive objects when we are able to finally recognize the sensory and motor patterns associated with them and, further, to anticipate the changes in our own sensorimotor activity as a result of perceiving (Di Paolo 2017, 19). These processes are what spawn our sensorimotor contingencies, which guide us through our daily lives imbued with perpetual activity. As the environment changes and offers new affordances for action, we select various sensorimotor contingencies to coordinate together—a kind of mutual accommodation—as a means of responding to these changes in accord with one's situational needs and desires (108). But it is important to recognize the extent to which sensorimotor contingencies are operative of the whole body-mind; affect is another form of movement and perceptual stimulation and so shape sensorimotor contingencies and motivate which combination to enact when enacting the world. Di Paolo et al. explain that emotions are emergent higher order cognitive constraints on the dynamics of muscular neural and autonomic processes that integrate the body's affect (177).

For one to be able to be efficaciously sensitive to environment changes to select the appropriate anticipated contingency coordination, one must have mastered the cognitive skills comprising sensorimotor contingencies.[5] Sensorimotor contingency mastery is the mastery of skills that aid us in refining and acquiring new sensorimotor responses to changes in our

[5] Enactivists critique functionalists, such as Noë, for positing sensorimotor contingency laws. They argue that the laws themselves rope in, and commit one to, unnecessary internal

environment; and, further, to then incorporate these new coordinations into our sensorimotor lexicon (37). Mastery is necessary for us to adapt when needed so as to be able to enact our world. They explain that mastery can be conceived of as a dialectic, world-involving perspective; world-involving interpretations constitute mastery. Importantly, they do not require internal representations—the cognitive processes of mastery draw on our procedural know-how instead (34). As if anticipating their pairing with Sheets-Johnstone, they argue:

> I live within activities that cluster sensorimotor schemes into functional and structural self-sustaining relations. Within these activities, I live in a situated normativity that shapes the space of my current affordances. Inside a cluster, schemes "make sense" in relation to each other and this permits not only the regulation of immediately adjacent consequences but of streams of virtual activity at the moment I enact a single particular schemes….these past, present, and future influences shape the enactment of the scheme here and now, and the enactment in turn modifies these virtual influences…we are thus frequently *thinking in movement*…. (231, italics mine)

This know-how—our embodied implicit knowledge—makes up our pre-reflective stance that helps us understand how, exactly, to go about enacting our world through the mastery of our sensorimotor contingencies. As such, the pre-reflective stance which is part and parcel of agency is a sensorimotor agency; agency is in the body as a combination of the codetermining organic body and lived body (174). Given the role of affect in our cognitive processes, disruptions to our sensorimotor agency gives rise to palpable emotional distress. Di Paolo et al. argue that this sensorimotor agency qua organic plus lived body is nonreductive insofar as it is an intersection between neurodynamics and personal experience. The agency of thinking through movement is felt through our intentional processes of framing, selecting, and enacting meaningful sensorimotor schemas (186). Agency is intentionally asserted during one's action plan that culminates in their desired outcome; agency is felt intentionally here because is exerts a feeling of control via a sense of achievement (191).

There are a number of components at play here that signify the distinctively normative nature of enactive sense-making. Generally speaking, the act of enacting meaningful sense in, and of, the world is the overarching

representations. But the notion of mastery that Di Paolo et al. are positing is not dependent upon the existence of sensorimotor contingency laws.

normative objective of a sensorimotor agent. Perception turns out to be inherently meaningful in terms of our relationality to our environment and other people. We go about selecting sensorimotor schemas by assessing and evaluating our relations to others and we do so by selecting those aspects of our circumstances that are meaningful to our actions vis-à-vis our needs or desires. All of these notable normative facets are bolstered by numerous other physiological normative conditions and constraints. That is, the normativity of lived body is grounded in the physiological, biological, neurological normative conditions for the biological and phenomenological body. These normative constraints for sense-making at the sensorimotor level include those listed above, such as adaptability, coordination and harmony, success and failure, sensitivity, but also: efficiency, robustness, adequacy, dexterity, elegance, coherence, and the affective, felt enjoyment of movement and flow of action (156). Further, because sense-making is inherently relational, the normativity of sensorimotor agency always contains a social component. The social norms associated with sensorimotor agency are, primarily, harmony, assimilation, and accommodation. These are the norms that guide individuals in their actions with others in their environment. For individuals to succeed in their action goals, they must use social cues to determine how to establish equilibrium between themselves and the others—people, places, things—involved in the affordances of the circumstance. As Di Paolo et al. rightly assert, we identify and describe people according to their behavior in, and through, action; this is normative phenomenological agency at the sensorimotor level (142). The normativity of sensorimotor agency brings this framework into relation with movement-knowings.

One way that physiological and social sensorimotor norms contribute to movement-knowings in dance is at the level of creativity in skill development. Enactive models of creativity move creative skill development away from internal processes and locates it in processes that stem from environmental and social sensorimotor schemas. Kirsh (2011) explains that the propositional knowledge involved in embodied cognition leads individuals to use their own sensory systems as simulation systems. This, then, allows individuals to use their own bodies as the active tool—the cognitive medium—for creative processes (1). When creating new dances, dancers must shift through, and across, their various sensory modalities to imagine new movement orders and transitions. I would argue that this is an enactive-friendly explanation of embodied cognitive creativity that draws on motor rehearsal across modalities to explicate the notion of curiousing. I say enactive-friendly because Kirsh's account of embodied cognition and creativity in dance making is a functionalist account. He argues that embodied cognition is a distinct form of computation distinct from symbol manipulation. The body and its sensory systems—especially the soma-sensory system—are used to simulate processes, where simulation leads to understanding (2). He insists that the use of other sensory modalities beyond motor modalities results in representations that may make it easier for the dancer to think through new movement affordances. I purport, however, that he seems to be using an idea of motor rehearsal to help explicate how dancers piece together new movements non-propositionally from their various sensory simulations. If so, this makes the core idea regarding embodied creativity in Kirsh relatively compatible with enactivism.

Rucińska and Aggerholm (2019) offer an explicitly enactive account of embodied creativity for the development of expert skill aptitudes and embodied decision-making in action. They specifically focus on tactical creativity, which is the creativity most related to sports. The cognitive processes related to sports are: divergent thinking, flexibility, novelty, convergent thinking, and evaluation; all of these cognitive processes are traditionally believed to be representational in nature. Instead, they argue, that creativity in skill execution stems from one's ability to develop aptitudes of responsiveness to available affordances as the situation plays out (675). Specifically, they argue that creative performance in sports is the creation of new bodily meanings that reveal or express new motor significance in the midst of one's situation (679). Enactive creativity, then, emerges when one has new embodied interactive experiences with their environment and is not merely the assemblage of disembodied ideas, which might be closer to what Kirsh's account suggests. Embodied creativity is a creative responsiveness to affordances (680). They also refer back to Gallagher about the importance of enactive reflection in action that I argue would be beneficial to phenomenological accounts of dancing. They say that embodied reflection is both a skill and an embodied coping. The reflection of the mover is not separated from them or their understanding of their movement in time and space—it is a part of it, a dimension of the flow. Knowing when and how to reflect in the act of skill execution and creation is a skillful way of interacting with the affordances made available through the new interactions with her environment (689).

One reason that Kish seems to be committed to a functionalist account of embodied creativity is because he highlights the centrality of the soma-sensory system in simulation for movement creation. However, Dunagan et al (219) argue that somatic practices are most consistent with the enactivist model of embodied cognition. They argue that contact improv shows an emergence of self-organizing behavior at the level of the body-mind because it draws on the automaticity of self-reflective observation in the moment (553). They rightly argue:

> Each individual's movement must arise from a combination of that person's sensory and conscious interactions, while simultaneously being shaped by the constraints of maintaining a continuously moving point of contact with a partner. In mutually allowing the point of contact and momentum to inform the direction of movement and the forms of weight-sharing that arise, both dancers place themselves in a situation in which they must give up some conscious decision-making so as to allow the bodymind to negotiate the changing terrain. Simultaneously, they must maintain an active awareness of what is happening and where the dance *might* go. Dancers must both allow their bodies to make choices based on the physics of any given moment, especially gravity and momentum, and be prepared to make conscious, deliberate choices on only a moment's notice should the need or opportunity to do so arise. (ibid.)

These enactive accounts of embodied creativity repeatedly substantiate the necessity of reflection and decision-making in dancing, which are paramount to practices of curiousing needed to engage in an embodied critical inquiry. Movement-knowings are fully dependent on one's ability to have this kind of creative responsiveness to affordances that open and close as the movement progresses and changes. Affordances made available through curiousing are fleeting and one must be able to have constant reflective awareness of where they are in time and place to creatively explore new opportunities for creative movement.

But there are more specific ways to understand the role of enactive cognition in somatic practices that undergird inquiring-in-movement. Movement-knowings of all kinds, not just those that create and experiment with novelty need a solid foundation of, and access to, embodied knowledge. And the dancer needs to be able to access and deploy—through procedural non-representational processes—embodied knowledges to tap into their soma. What is needed, then, for an enactive framework for cognitive critical inquiry that relies on access to our embodied knowledge through somatic work is one that sees the body as linguistically communicative all the way down. This foundation comes from Di Paolo et al.'s (2018) account of the linguistic body.

According to Di Paolo et al., persons have a linguistic agency that is grounded in a linguistic body. This is why enactivism is a cross between sensorimotor sense-making and participatory sense-making. As sensorimotor agents who are in a dynamical system with our environment, this puts us in a constant readiness to act; we exist in relation to the world in a participatory readiness to engage others through enactive practices. When we think about how language is embodied, we must think about how our bodies language. An enactive theory of embodied languaging practices places the focus on our basic corporeal sensorimotor logics (133). Our linguistic relations to the world emerge from our social interactions with others, which provide the parameters on goal construction and action affordances. Thus, a linguistic body is a body that is socially situated and performed and from which others' agency can be incarnated (195). We have a linguistic agency, that is a sensorimotor agency, because we must incorporate the acts and utterances—embodied and verbal—of others into our assessment of opportunities for action. Utterances, as a communicative medium, entail the relationships between people and acts. They astutely avow that "[t]o incorporate an utterance, in consequence, is in part to *incarnate* others" (193). Our agency necessarily depends on the kinds of goals others have as well, since their cares and concerns contribute to, and shape, my own choices. Linguistic bodies are bodies that navigate the constitutive tension between how I choose to act and how I incorporate other's agency into my own (217). Therefore, all of our actions in the world are part of a shared relationality and consists of the coregulation between oneself and others that grounds participatory sense-making. They explain that interactive situations of participatory sense-making, even when they are not recognized as such, are the entry points into the categories that develop and describe a deeply embodied approach to languaging (137). We develop and enact cognitive powers via linguistic agency through social practices of coregulation in sense-making. While we have autonomous control over how we make sense of the world, our ability to make sense requires that we must allow for others to engage with us as linguistic bodies in joint social acts (149). This is a reflective social form of agency and sense-making (185).

The enactive view of sense-making is social and dialectical in nature precisely because it is dynamic and embodied. Both dynamical systems and dialectical practices are constituted by norms such as circularity, narrative, relations, tendencies, non-linearity, and emergence. Thinking about the world dialectically, they claim, just is a means of thinking about the world enactively (109). The dialectical, dynamic components of participatory sense-making reveal how we make ourselves by making ourselves out of others as perpetual bodies in action. Consequently:

> In the enactive view, in other words, an authentic human self is one that understands her own constitutive dependencies and her own incompletion, and in so doing contributes creatively to—and struggles critically against—the framing that community processes impose on herself and her people. (257)

The creative responsiveness of enactivism also drives linguistic bodies to be responsive to one another in ways that bolster autonomy and secure relationality. How we choose to incorporate others' actions is a function of how we perceive their actions in relation to our own desires. They exist, in part, as an intentionality we have "over there". Intersubjectivity, then, rests on a synergy between bodies in daily encounters in the world (63). Sense-making is intersubjective and participatory sense-making evolves from coordinated interaction patterns between the self and other, even if not noticed and even if outright rejected. No matter the explicit desire of an actor, patterns of social interaction produce entanglements between the bodies of participants that create and refine social meanings and understandings (77). They explain further:

> The incorporated flows of utterances that make up a linguistic agent are *always* the joint result of personal enactments and of patterns that live in the community; linguistic bodies—the embodiment of linguistic agency—are *both* personal and constitutively social. They are communal achievements. (193)

The joint social practices of frame building and sense-making involve our expertise in ethical know-how that follows from being a linguistic body. Our implicit know-how resultant of interactive relationality imbues the acts we perform as linguistic bodies with this normativity. Our social powers qua linguistic bodies is our ability to bring into effect, or question, norms and attitudes. Linguistic performances always carry ethical content since linguistic performances always carry deeply embodied risks in how we incorporate them into our sense of self and how we incarnate others (314).

Linguistic bodies contribute partial acts and utterances to the grammar of embodied, participatory sense-making. Our linguistic agency consists of making these embodied utterances coherent. Di Paolo et al. define utterances as acts—embodied, material patterns—enacted by sensorimotor bodies. They have an inherently dialogic structure that orients people towards each other; uttering in interaction is taking one's turn in the interaction, as it were, and opens the space for meaningful responses. As such, they engender the necessary social meaning-making norms of recognition and interpretation. "They are acts that demand, and create, participants" (2). Partial acts are partial because they are initiated by one actor and must be completed by the other in order to create a whole act (154): high fiving, carrying stuff, holding an elevator, sex. They are used by actors who—knowingly or not—work together to codefine and coenact mutually defined sensorimotor schemes to facilitate coregulating enactments. Each newly developed and enacted schema contributes to our embodied repertoire of schemas for interaction; if we have a large repertoire of partial acts available to choose from during interactions, they can be chosen to start new interactions and potentially spawn newer modes of coordination. I may introduce my hand as a partial act of greeting but this creates a new opportunity to contribute a wholly new and unanticipated greeting return, since my open hand allows for plenty of affordances for completion. For this reason, gestures—like sticking one's hand out as a greeting—function as regulating partial acts to help interactions sustain coregulation by making choices amongst affordances more obvious and easier qua social repetition. This ease provided by gestures shores up our social know-how with respect to interaction.

One of the reasons why their account of linguistic bodies seems so pertinent to my project is in how they construe movements and gestures as utterances. That partial acts within interactions function as linguistic contributions in a deep, rich, and technical sense helps one better understand how reading bodies is sensical. Enactivism allows embodied cognition to go further than the cognitive linguistics of Lakoff and Johnson.[6] Sure, our language is amply constituted by, and through embodied metaphors. But

[6]The account of linguistic bodies is significantly more robust than what I supply here; my aim is to provide just enough information relevant to meaning-making at the embodied level so that I can connect to movement-knowings. In the book, they extend the language and grammar of the body to the language and grammar of verbal language, as well as social interactions based on shared group status and histories.

thinking of thinking-in-movement and thinking of bodies as linguistic bodies reveals that the relationship between the body and language also works the other direction. That is, the body yields dynamic and thick meanings and verbal language is a sort of fast pass to getting at that meaning. Verbal language fails to capture all of the dimensions of lived life because it is impossible for a static medium to capture moving meanings and movement-knowings. The sensorimotor grammar of the body articulated by Di Paolo et al.'s framework clarifies how moving meanings and movement-knowings are not merely accentuations or ornamentations for emphasis in propositions.

The dynamic and dialectical nature of sense-making by sensorimotor bodies helps clarify how somatic exercises access movement-knowings.[7] Bodies develop movement-knowings through their interaction with their environment and through interactions with other linguistic bodies. The thick meanings that emerge from sensorimotor contingencies and mutually accommodating sensorimotor schemas become embedded into the body and can be enacted at later moments when one attempts to impart sense onto their environment or in interactions with others. The meanings come in through action-perception in interaction and they are sent out again in the same way. This is a simplified articulation of building movement-knowings through enactive dialectical exchanges. There is always an excess—or rather abundance—of meaning held within the body that spills over from verbal communication, and even partial acts. This is because partial acts require completion by another, and it is not uncommon for partial meanings to be read and interpreted by the other incompletely, even if not inadequately. All this is to say that there is a deep well of meaning within us ready to be enacted. This is the source of our movement-knowings. If we regard somatic practices as naturalized phenomenological endeavors—rather than as hippie-dippie exploits—we recognize them as practices that stimulate active enactive self-reflection that connects to our affective perceptual capabilities. And, more importantly, training in somatic exercises is a means to train our self-reflection to make stronger and more penetrating connections to our sensorimotor affective concepts. Active self-reflection is a way of looking inward; but instead of

[7] See also: Sharon Chaiklin. "We Dance from the Moment Our Feet Touch the Earth". In *The Art and Science of Dance/Movement Therapy: Life is Dance*. Chaiklin, Sharon and Hilda Wengrower (eds.). New York: Routledge Press, 2016: 3–11.

looking to find visual or verbal knowledge, it is able to connect to the sensorimotor patterns picked out by embodied arousal to the subject of inquiry.

More specifically, with respect to somatic exercises that generate connections to embodied knowledge through movement and arousal monitoring, I am thinking of the sensorimotor habitat and sensorimotor strategies. Buhmann et al. define the sensorimotor habitat as sensorimotor contingencies that are created through specific patterns of agent-environment interactions. They define sensorimotor strategies normatively and epistemologically—sensorimotor coordination patterns are regularly used because they have been evaluated as preferrable to achieving a goal. I think of somatic training as closest to sensorimotor strategies because the acquisition of sensorimotor strategies describes how agents become attuned to a specific situation, by selecting and modulating sensorimotor coordination patterns in accordance with relevant norms. "The notion corresponds to an even more agent-centered and history-dependent idea of SMCs and it seems close to some of the uses of the term that link SMCs and personal level phenomena". (Buhmann et al. 2013, 4, 5–6)

Another interesting point that the account of linguistic bodies raises for movement-knowings is with respect to reading one's own body. When one is engaged in somatic-based improvisational movement, the body is—in a sense, or in every sense—in dialogical exchange with itself. Improvisation can be best understood as a mode of self-directed utterance, which would make the process of an embodied critical inquiry a reflexive mode of social cognition. When one engages in a self-directed exchange with oneself, there is a sense in which the self is in a participatory sense-making relationship with itself. Di Paolo et al. tell us that incarnation is entailed by the possibility of the flow of self-directed utterances since they induce a role distinction between the speaker and the audience. If the person is purposefully engaged in a critical inquiry within themselves, this positions themselves in this relationship to themselves. Thus, the self-directed flow, specifically the bodied movement accessing embodied knowledge, establishes a dialogic attitude towards oneself (193). The mover incarnates herself as interlocuter through enacting her bodied movement-knowings. This movement can be read by herself, as her interlocuter, through the various linguistic sensorimotor affective patterns she pushes to the surface. With respect to the relationship between syntax and action, they explain:

> Just as linguistic bodies are grounded on the sensorimotor dimension of embodiment, so the shared grammatical know-how for structuring utterances into sequences...helps to shape novel forms of ***sensorimotor*** sense-making...The embodied know-how for conjugating complex sequences of sensorimotor schemes is learning in participation. Controlling complex action is one of the well-known effects of self-directed speech...and the effects of linguistic categories and dialogic attitudes on decentered perception are well established.... (291–292)

In the case of an embodied critical inquiry, this manifests through somatic-based improvisation qua self-directed movement-knowing utterances.

And in just the same manner, movement partners are able to have successful dialogical exchanges through contact improvisation. Likewise, the embodied shared know-how of participatory sense-making in social interactions pertains directly to the practicability of embodied dialogues in partner work.[8] The same relationship between sensorimotor contingency syntax and action clarifies how partners engage in intelligible moving discourse together. What is different in somatic-based partner work is that the discursive movement exchange is simultaneously self-directed utterance and other-directed utterance since the partners move as a part of the other.

Dancing Metaphors and Enactivism

This theory turns on the compatibility between embodied metaphors and an enactive framework. Certainly, it is possible for the embodied metaphor-based method itself to succeed outside of any properly articulated grounding framework—since it is successful—but the goal is less to yield the method than it is to provide a philosophical analysis of how it is possible to use movement to ask oneself critical questions about oneself and the world. If I can shift the understanding of embodied metaphors from the domain of cognitive linguistics to enactivism, I can offer a fully embodied, fully sensorimotor, fully movement based, online reflection model of thinking and inquiring-in-movement. I find it odd that embodied metaphors are largely sequestered to the domain of cognitive linguistics since the whole idea is to illuminate the non-metaphorical way in which the

[8] See also: De Jaegher and Di Paolo (2007).

body knows more—more richly and just generally more—than static language depicts. I do think that it is important to remind the reader that theorists such as Lakoff, Johnson, and Noë identify as non-representationalist, which would immediately make the problem I raise here moot. But there are some enactivists who question the self-identification of these embodied cognitive theorists. So while I do accept their self-designations, I do want to attempt to explicitly appease skeptics about the pairing I consider to be necessary to the embodied critical inquiry framework. Here, I deploy Katan-Schmidt's notion of dancing metaphors to substantiate the congruency of embodied metaphors with the enactive model. But it may very well be that dancing metaphors are not quite the same as embodied metaphors since dancing metaphors are online and in action in a stronger and more robust sense than linguistic embodied metaphors might be.

The relationship between syntax and action in dialogical exchange helps explicate how embodied metaphors can be concomitant with enactivism instead of being limited to embodied functionalism. Because creative movements constitute sensorimotor embodied metaphors, their syntactical sequences constitute the larger narrative claims generated by, and through, the movement-knowings. This embodied, phenomenological process is consistent with enactivism because much of the sensual qualia that are explored for moving-meaning stem from affect. Katan-Schmid rightly purports that emotions are themselves embodied ideas, particularly when they are interpreted through and with enactive perception. When emotions are the working material for dancing and choreographing, they are enactive and enact perception. She argues that they are enactive because their evaluative properties give rise to considerations of the embodied creative possibilities and directions for further action in terms of movement development (Katan-Schmid 2017, 115). She rightly espouses that the physical movements of dancing are expressive of meaning by virtue of the fact that "…dance gestures are living forms. As an emergence by a living process, the gesture contains its process of inventive creation as an expression" insofar as they "reveal the thoughtful process of their shaping" (2016, 11). Katan-Schmid purports that dancing metaphors emerge from metaphorical interplay while exploring movement through dancing. This process requires developing established embodied and dance sequential movements into new actualizations of dancing concepts through the affordances generated through the utilization of "as if" rules within the bounds of kinesthetic laws—imaginary and actual. Because the logic of dancing permits of delimited but broad possibilities for future communicative movement—dancing from what is to what shall become—thinking through movement—enacts and realizes those possibilities as innovative physical ideas and bodily imagination (2017, 279).

From this, I want to take Katan-Schmidt's argument a bit further. The process of "as if" enactivism bids the dancer-choreographer to movement-knowings. Yet this process is not one of make believe or pretend; the dancer is not pretending to say such ideas with her body, she does say such ideas with her body through the newly begotten movement sequences and gestures. I reckon that to emphasize how dancing metaphors are actual rather than pretend is emblematic of, and gives substance to, her claim that dancing gestures are living forms. Given that embodied metaphors can be purposefully shaped, displayed, and communicated as truthful knowledge is why embodied metaphors can be non-representationalist over and beyond action-oriented representations. I maintain that performed bodied embodied metaphors transcend the notion of "as if" and constitute the notion of a dancing "is". Raymond Gibbs highlights the importance of embodied metaphor to dance as a means of communicative action by identifying how dance patterns convey shared metaphors are central to narrative uptake. Gibbs focuses on contact improvisation because he believes that this style of dance is highly efficacious communicatively since the emergent dance patterns and movements from spontaneous bodily interplay and response give rise to embodied conceptual metaphors of our most ordinary embodied experiences (2003, 187). He purports that dancing sets forth impressions of stories that can be understood through the tacit comprehension of shared embodied metaphors that, often times, neither the dancers nor the audiences are entirely aware that they are sharing. Yet they succeed anyway—at least thematically—because people, typically of a shared culture, have this tacit understanding of the metaphorical signification of abstract concepts through embodied experiences (192). Performed bodied embodied metaphors are directly communicative and are instances of direct cognitive processes. Shared sensorimotor contingencies might be the most direct manner in which to validate the ability of various individuals to share a sense of understanding when it comes to reading bodies.

More important to the task here is explicating how dancing metaphors helps individuals read their own bodies when their work is improvisational rather than fabricated with mostly cliché primary metaphors. Improvisational movement is going to be comprised of unpredicted and highly dynamic movement—not necessarily dynamic in the intensity sense, but dynamic in the sense that it will have many axes of movement dynamic intersections. I believe that Gallagher's and Lindgren's notion of affordances qua enactive metaphors are helpful here too. They define enactive metaphors somewhat differently than embodied metaphors though they are based on embodied metaphors. They explain that enactive metaphors are a way of engaging metaphors by enacting them and putting them into action or bringing them into existence through action (Gallagher and Lindgren 2015, 392). I purport that they are instantiations of embodied active imagination specifically in cases of thinking-in-movement. Their very essence is to support the thinking process and aid understanding. The most significant claim they make that I think shows how "as if" can help us read our own wandering/wondering bodies is that "[a] good metaphor will lead us somewhere, open up an insight, show us something that we could not see before..." (391). If it turns out that we make sense of the world, and come to understand the world, through our bodily processes in action—through thinking-in-movement—it seems to follow that we should be able to make sense of our own bodies in movement, specifically in terms of the dancing metaphors they enact, because our bodies already contain our knowledges and are explicitly expressing it through meaningful creative cognitive movement. When we read bodies for their embodied metaphors, we are finding the affordance for meaning and action in—and on—them. Dancing is a way of enacting metaphors that goes beyond the pretend play of "as if" and, again, takes us to the cognitive domain of simply "is".

The possibility of objectivity with respect to dancing metaphors—their readability—rests in the ability of the phenomenology of dance to be incorporated into a naturalized phenomenology. Dance, as I have shown throughout, continues to be taken up more and more within cognitive science. Accordingly, the phenomenology of dance is not only an art; it is now recognized as a subdomain for science. The advances cognitive scientists and neuropsychologists have made through the use of fMRIs for understanding what and how dancing can make-sense and have meaning is unfurling at record speeds. Calvo-Merino et al. (2005) have been using fMRIs to examine brain processes related to motor simulation and mirror neurons by using dance as the action simulation. As scientists continue to fold dancing into their research, their research continues to naturalize what was once considered almost strictly virtual or subjective. As research continues, my hope is that these sciences can work in tandem with philosophy to push dance and dancing to the fore of cognitive accounts of sense-making.

Bibliography

Albright, Ann Cooper. *Choreographing Difference: The Body and Identity in Contemporary Dance*. Hanover: Wesleyan University Press, 1997.

Best, David. "Dance Before You Think". In *Dance Education and Philosophy*. Graham McFee (ed.). Chelsea School Research Centre Edition 7. Oxford: Meyer and Meyer Sport, 1999.

Buhmann, Thomas, Ezequiel Alejandro Di Paolo, and Xabier Barandiaran. "A Dynamical Systems Account of Sensorimotor Contingencies". *Frontiers in Psychology* 4 (2013): 1–19.

Calvo-Merino, B., D. E. Glaser, J. Grèzes, and R. E. Passingham. "Action Observation and Acquired Motor Skills: An fMRI Study with Expert Dancers". *Cerebral Cortex* 15 (2005): 1243–1249.

Chaiklin, Sharon. "We Dance from the Moment Our Feet Touch the Earth". In *The Art and Science of Dance/Movement Therapy: Life is Dance*. Chaiklin, Sharon and Hilda Wengrower (eds.). New York: Routledge Press, 2016: 3–11.

Clark, Andy. *Being There: Putting Brain, Body, and World Together Again*. Boston: MIT Press, 1997.

De Jaegher, Hanne and Ezequiel Di Paolo. "Participatory Sense-Making: An Enactive Approach to Social Cognition". *Phenomenology of Cognitive Science* 6 (2007): 485–507.

Di Paolo, Ezequiel, Thomas Buhrmann, and Xabier Barandiaran. *Sensorimotor Life: An Enactive Proposal*. Oxford: Oxford University Press, 2017.

Di Paolo, Ezequiel, Elena Clare Cuffari, and Hanne De Jaegher. *Linguistic Bodies: The Continuity between Life and Language*. Cambridge: MIT Press, 2018.

Dunagan, Colleen, Roxane Fenton, and Evan Dorn. "Modelling Improvisation as Emergence: A Critical Investigation of the Practice of Cognition. Life Practices". In *Oxford Handbook of Improvisation in Dance*. Midgelow, Vida (ed.). New York: Oxford University Press, 2019: 545–562.

Fraleigh, Sondra Horton. *Dance and the Lived Body: A Descriptive Aesthetics*. Pittsburg: University of Pittsburg Press, 1987.

Gallagher, Shaun, and Dan Zahavi. *The Phenomenological Mind*. New York: Routledge Press, 2012.

Gallagher, Shaun, and Robb Lindgren. "Enactive Metaphors: Learning Through Full-Body Engagement". *Educational Psychological Review* 27 (2015): 391–404.

Gallagher, Shaun, and Zuzanna Rucińska. "Prospecting Performance: Rehearsal and the Nature of Imagination". *Synthese* (2021): 1–19.

Gibbs, Raymond W. Jr. "Embodied Meanings: Performing, Interpreting, and Talking about Dance Improvisation". In *Taken By Surprise: A Dance Improvisation Reader*. Albright, Ann Cooper and David Gere (eds.). Middletown: Weslyan University Press, 2003: 185–194.

Katan-Schmidt, Einav. *Embodied Philosophy in Dance: Gaga and Ohad Naharin's Movement Research*. London: Palgrave Macmillan, 2016.

———. "Dancing Metaphors; Creative Thinking within Bodily Movements". *Proceedings of the European Society for Aesthetics* 9 (2017): 275–290.

Kirsh, David. "Creative Cognition in Choreography". *Proceedings of the 2nd International Conference on Computational Creativity* (2011). n.p.

Marin, John. From *The Dance*. In *What Is Dance?*. Copeland, Roger and Marchall Cohen (eds.). New York: Oxford University Press, 1983: 22–23.

McKechnie, Shirley. "Dancing Memes, Minds and Designs". In *Thinking in Four Dimensions: Creativity and Cognition in Contemporary Dance*. Grove, Robin, Catherin Stevens, and Shirley McKechnie (eds.). Melbourne: Melbourne University Press, 2005: 81–94.

Nahrstedt, Klara, Ruzena Bajcsy, Lisa Wymore, Renata Sheppard, and Katherine Mezur. "Computational Model of Human Creativity in Dance Choreography". Association for the Advancement of Artificial Intelligence, 2007: n.p.

Rucińska, Zuzanna, and Kenneth Aggerholm. "Embodied and Enactive Creativity in Sports". In *Handbook of Embodied Cognition and Sport Psychology*. Cappuccio, Massimiliano (ed.). Cambridge: MIT Press, 2019: 667–694.

Sheets-Johnstone, Maxine. *The Corporeal Turn: An Interdisciplinary Reader*. Charlottesville: Imprint Academic, 2009.

———. *The Phenomenology of Dance*. Philadelphia: Temple University Press, 2015.

Stewart, John, Oliver Gapenne, and Ezequiel Di Paolo. *Enaction: Toward a New Paradigm for Cognitive Science*. Boston: MIT Press, 2010.

CHAPTER 7

Embodied Critical Inquiry: A Method

As I discussed in the introduction, my motivation for developing this account came from my experience with the circus crowd during some of the more politically active moments of the pandemic and Black Lives Matter protests. I had seen previously, but was beginning to see more frequently, attempts made by artists to address social injustices through performance art. Sometimes, shows would have social justice themes, sometimes artists would put something together on their own that connected to social justice issues. But my reaction to these admirable attempts to meld circus with justice, which is actually a long-standing relationship, was soured by the glibness some of these shows and acts contained. I understood and recognized the deeply authentic desires of these artists to speak to social injustice, but it didn't appear as if much preparation or thought went into these pieces.

The philosopher in me wants to insist on more affective and intellectual investment on behalf of performing artists—not only with respect to social justice issues, but just with respect to any issue at all. When artists want to address concepts like beauty, I want them to think about why they want to address this issue. I want them to think about what beauty even means. I want them to question their own relationship to beauty. In short, I want performing artists to be philosophers in motion. I do not mean that creative artists are not philosophers. Quite the contrary. I

S. Welch, *Choreography as Embodied Critical Inquiry*,
https://doi.org/10.1007/978-3-030-93495-8_7

believe that creative artists are inherently philosophers. My goal for developing this method, then, was to construct a process through which performing artists could be more intentional with their choreographing engagements. Certainly, many artists already have a sophisticated process they follow, others may have a strongly intuitive procedure, and some may have no desire at all to choreography in such a precise manner about such precise ideas. But there are some who would like to. And there are many instances, such as in cases where artists are attempting to breach topics related to oppression, where it might become a bit more obligatory to engage in an explicitly reflective process prior to creating a public art piece. This method is but one way that artists could choose to do so.

Deciding how to go about formalizing this method was initially overwhelming. Because I do not engage in empirical work, I didn't even know where to start. What I did have was a felt sense of what this process should be and how this process would pan out. After much contemplation, and much begging for funds, I finally determined that the best way to present this method as philosophically valid would be to actually conduct the experiments needed to substantiate my claims. I chose four performing artists to go through the method. The first two are aerialists: one was a professional ballerina and the other is a full-time professional performer. I work very closely with these two artists and felt that they each had the right amount of knowledge to help me work out the kinks and the right kind of personal relationship to me to be willing to go through something entirely experimental. They had enough trust to let me lead them through what turns out to be a highly emotional process and I trusted them to give me unfettered criticism and corrections. Without these two artists, this method would not have been developed to the extent that it is. I did not have personal relationships with the other two artists. One artist was recommended to me and is actually an academic Psychologist who is partnered with a Philosopher, so she was ready for madness. The other is a classically trained full-time dancer and dance teacher with whom I've crossed paths. Each of these artists, and their varying skills, training, and connections to me, each provided unbelievably indispensable feedback for the formalization of the method. I will go over each of their experiences in the process in the next chapter, which covers the empirical results from the experiments.

Somatic Exercises

In Chap. 4 I posited the priority of somatic work for an embodied critical inquiry. This claim is in no way overstated. Given that I do not personally have any training in somatic work, I had to resort to doing a lot of independent research to learn about various methods and various ways of going about it. While Fraleigh's book on somatic work was probably the most significant for the philosophical discussion, I would argue that the dance movement therapy books were most beneficial for providing somatic substance for the method. There were a few books from which I drew examples—some of which I used directly and some of which I adjusted or blended with others—the two most significant being: *Dance and Somatics: Mind-Body Principles of Teaching and Performance* (Brodie and Lobel 2012) and *The Art and Science of Dance/Movement Therapy: Life is Dance* (Chaiklin and Wengrower 2016). In this section, I lay out some of the specific somatic exercises and corollary key points that I found to be extremely valuable so as to position the role of somatic work at the fore of the process.

Somatic work can manifest in varying ways depending on the method deployed or approach taken by the witness/facilitator. Given that my intention is to use somatic work to access deep body knowledge through movement, I use somatic exercises as a sort of "priming". That is, I facilitated the mover in varying somatic exercises over a long time. While some facilitators incorporate more explicit modes of movement during somatic work, I mostly delimited my own exercises to static, prone somatic work to help the mover direct their focus and attention to the chatter of the body. From the outside, the somatic exercises I use would easily resemble modes of meditation given the stillness of the mover—especially when the mover begins to prime herself. But instead of aiming for stillness of the mind, the goal is to raise the internal rhythms of the body to the surface; the somatic work I employ is very far from the idea of stillness but could be considered a form of turbulent mindfulness. More specifically, the overarching goal is to direct the mover in turning inward to locate, and track, her embodied intuition by helping her focus her perception. A secondary goal is to teach the mover how to facilitate herself so that she can accomplish the process on her own.

The focus and direction of the mover's perception was the most difficult and crucial component of the somatic work. I spent roughly 30 minutes on this somatic priming section of the process. The facilitator must be able to execute subtle and limited verbal cues to help the mover move her perception in extraordinarily specific directions. Rules for effective facilitation that I learned over time: do not rely on imagery or metaphorical language, do not use "do not" cues since this will immediately cause it to happen, do not move swiftly from one cue to the next because the mover will not be able to focus and the exercise will fail, do not over cue because this will overwhelm and impede the focus of the mover, make sure to ask if the mover has any discomfort in their body and to inform them that they should find a position that will not aggravate this discomfort, ask the mover how much direction they prefer to have and let them tell you when to adjust along the way, and inform the mover that they can choose to have their eyes open or closed but remind them that they will want to at least open their eyes somewhat during movement exercises.

First, the mover must be able to identify and isolate each of her senses and identify patterns and rhythms. Then she must be able to find ways of combining various pairs of senses according to a discovered shared pattern and/or rhythm. One example I used includes picking out and tracking each of the noises and finding a shared pattern and then matching the breathing to this construed pattern. It is important to note that this process need not take place in silence because silence is often times louder than any other context. The trick is to help the mover realize that all noises are simply part of their lived experience that contribute to, and shape, the wide-ranging forms of knowledge that go into the body. Rather than find certain noises to be distractions or interruptions, the mover should incorporate each noise into the symphony. We were even able to successfully complete these processes in a gym that had an ongoing children's camp. Where difficulties arose was when unexpected noises that did not belong to the context entered. But the largest difficulty was going through the process in silence because every noise, such as a surprise thunder strike (which happened) is unavoidably jarring.

Other sensory stimuli I directed the mover towards were textures and how their kinesthesia changed when thinking of them as touching her versus her touching them (workout clothes were a big winner here), temperatures of different parts of the body, and light patterns on the inside of the eyelids. The specificity of cues towards unexpected or unnoticed sensory input was most effective in turning the mover's perception inward to connect to the body-mind and, accordingly, embodied knowledge.

There was a specific order of perception direction that seemed to work well across all of the movers in each of the sessions. First, I took the mover through the various sensory modalities to identify the various rhythms. I did not direct the mover to focus on their breathing or to moderate their breathing. I merely suggested they take note of their breathing rhythm and compare it to their perception of the rhythms of the sensory stimuli. I then directed the mover to seek out the kinesthesia of their embodied intuition and to locate where they kept it in their body. I originally referred to this as the "basket" of their intuition, but this was too visual, so I changed it to the "place" of intuition.

The second step in the static somatic priming was the center-periphery exercise. I would direct the mover to track those same sensations but through moving perception. That is, I would instruct them to gather the rhythms of the sensory stimuli into the place of their intuition and then push it out to the periphery of their body from their center (this is usually where people found their intuition). The exercise is a slow repetition of leading the mover through the process of purposefully moving their sensory perception around in a directed manner. One trick I learned was that it was useful to include cues regarding the extreme peripheries of their body, such as the knot on the back of their necklace or the place where their bangs touched their eyebrows. Because there is so much sensation and so many points—infinite points—of body-environment contact, offering cues of places one would never think of works well to keep the mover in a focused state and in tune with the depth of their body's contact with the world. The third step is similar to the center-periphery except the focus is on the layers of the body. What I found to be interesting is that each mover conceived of their body layers differently. One person felt her layers to be concentric circles, and another felt her layers to include the ground below her and the air above her.

The second stage of somatic work is (minimally) dynamic. I would direct them to move from prone, to sitting, and eventually to allow the rhythm(s) of their intuition to move them to standing. They were then directed to repeat the center-periphery and layers exercises but in a way that allowed them to shift and be ambulatory. So as the body attended to itself in one direction more heavily than another, they would balance more in that direction. Because I was using professional movers, they used this opportunity to really get into the edges of their balancing and extension without actually moving beyond the intention of minimal movement. The idea was to let the felt body sense push to the surface and allow it to do so and not to engage in actual "movement", which was reserved for the improvisational exercises.

The final stage of the somatic priming work is a modified version of Kelly Ferris Lester's memory-based Dance Map (Fraleigh 2015, 222), which I just call the "dancing memory" exercise. This exercise went through many iterations, and it still changes when I do it; there's a feel for what is needed in the moment for the exercise to take hold, which just generally holds for all of the facilitating. It also depends on what stage of the training process it occurs in and how quickly the mover is picking up the somatic skills. My goal for this exercise is to help the mover figure out how to take remembered sensory perception, such as that tied to body knowledge, and to actualize it on the surface of the body. At first, I was very skeptical about this exercise but when I saw the Dance Map exercise, something just told me that I had to use this. I also assumed the movers would find it bizarre and entirely unrelatable, much less effective; in short, I thought they would think it was trash and lose all faith in me. But it turns out, this was an incredibly valuable exercise that the movers not only felt themselves to be successful in but actually enjoyed the most of all the exercises we did throughout the process. I learned not to conduct this somatic exercise until after the concept was introduced. But sometimes the order of somatic exercises change since the order of static position to improvisational movement needs to be in a certain order.

Once the mover was sufficiently somatically primed in relation to the concept, I would ask her to select one memory. I would say that it did not have to be the first memory that came to mind but that it also should not be a specifically sought out memory; the memory they should select should be the one memory that jumps out to petition for attention. The next step would be to instruct the mover to play all the way through the memory from beginning to end in real time but to be sure to account for the specifics of the background context for the memory. Then I would instruct the mover to play through the memory in slow motion a few times to pay attention to different aspects of the memory in each review. The fourth step is to instruct the mover to follow a similar somatic priming process regarding the senses except that they should be tracking each sensation in the memory. I learned that it was important not to make too many cueing suggestions because if I indicated that there might be someone else there to account for, but they were alone in the memory, they would quickly build someone else into the memory. So minimal and vague sensory cueing is imperative. Next, I would instruct her to bring those sensations and rhythms to the body here and to feel the sensations "as if".

This is the point in which the mover would be working to bring the content of their embodied knowledge to the surface to create a strong connection to their body-mind beyond the present.

The part of this exercise that leads me to refer to it as the dancing memory exercise follows from the sensory recollection process. Because the goal is to bring the mover into movement, and because sensory rhythms are so important to accessing movement-knowings, I instruct them to focus very specifically on the movement within the memory. This does not only pertain to the movement of persons. The movement could have been the smell, the fabrics, the sounds. The goal is to, again, have the mover establish a series of rhythms and pull them into, and onto, her body. Next, I would ask the mover to allow the movement of the memory to take her to a standing position in step with that movement. The next few steps would lead the mover into their improvisational movement. Then, I would instruct the mover into imagining the movement of the memory leading her in movement as if the memory were a dance partner. The final step would be to instruct the mover to imagine stepping into the memory-partner and begin to dance the memory and the meaning of the concept. This would lead directly into the improvisational movement.

The improvisational movement exercises are driven by the somatic state the mover takes on. The idea of "dancing the concept" is intended to direct the mover through her movement. More specifically, the use of somatic priming is intended to make it the case that the concept moves the mover rather than the mover acting out the concept. The mover should follow the somatically generated felt sense and affects of their embodied knowledge as they are felt rather than attempting to articulate an explicit idea and emotion. The mover should be encouraged to pay attention to thoughts, emotions, and memories and allow these experiences to move the body in any way that feels right with respect to their meaning of the concept—their personal connection to it and, thus, what it means for them. This process allows the mover to engage their embodied active imagination to reveal their conceptual movement-meanings.

If I've learned one very important thing during this experiment, it is that people can dive very deeply into their bodies, and they all hear and feel their movement-knowings in very particular and idiosyncratic ways. It was beautiful (Fig. 7.1).

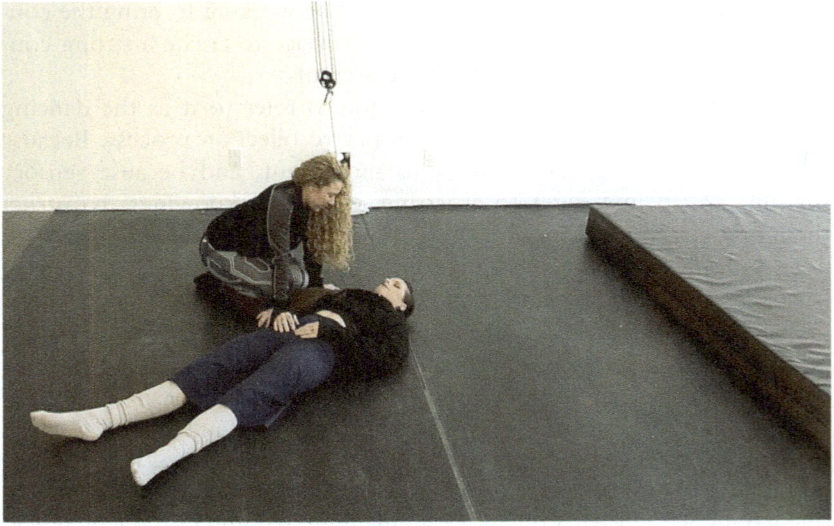

Fig. 7.1 Constance doing somatic exercises

Process: Inquiry to Narrative

It is difficult to determine which should come first—the method explanations or the method outline. It seems as if it would be preferrable to start with the explanations because then that would help clarify some of the steps for the method itself. However, the explanations don't make sense in the absence of the method steps. Therefore, I have chosen to put forward the method itself and offer key practical points for some of the steps to indicate their purpose or how they work. Generally, the method should be straightforward. But having the method first provides context for the remarks I provide on it shortly thereafter. Because the method involves worksheets, I provide those in the next section. This allows the reader to have all practical material at hand. And since the method involves a learning and training curve for the facilitator and/or mover, I include both the beginning version of the method and the ideal form of the method once all exercises and orders are picked up. Again, it is important to note that this is just one version of an embodied critical inquiry method. The exercises are open to adjustment and the order can be adapted. The reason I present the method in this specific fashion is because these exercises and this particular ordering proved to be most efficacious upon the conclusion of the four experimental periods.

I follow the embodied critical inquiry method with a short version of the corollary embodied metaphor narrative construction choreography method. This, too, can be adapted. I include it here because, ideally, the mover will be able to go through a choreography process from beginning to end using a unified method. I imagine the cycle going something like: an idea gets introduced, such as when companies or troupes want to produce a show, then the mover can engage in the embodied critical inquiry process—using embodied metaphor movement analysis—to articulate specific questions they may have about the theme and want to explore. Once the mover completes the embodied critical inquiry method, they can then ruminate on the question to figure out how best to answer it, which would be the content of the narrative choreography. Finally, the mover would construct and blend various embodied metaphors that best communicate the content of their narrative. Many of the details will unfortunately remain somewhat vague until the next chapter, where I set out each mover's experience with the method and their completed (or incomplete) worksheets.

To some, this method may seem overly rigid and will not feel appropriate or conducive to their own approaches to creativity. That is to be expected. This method is not for everyone. This method is well suited to people who are overly organized and are inherently resistant to ambient free flow, for people who do not have a lot of experience choreographing from a free flow state, for people who are learning important things to think about when creating from an authentic personal place, for people who are stuck and aren't really sure what they think about a particular idea, etc. My training partner hates the whole idea of structured creativity while I can't create without systemized structure. Needless to say, our creation process can be challenging. This method is an aid for people who need to structure and organize their ideas. Creativity comes in all forms so it is helpful for many modes of creation to be available so that more people can participate in art.

Beginning Inquiry Method

1. Static somatic exercises: sensory isolations, rhythm tracking, center-periphery, layers, locating seat of intuition
2. Generic dancing memory exercise: not concept specific
3. Ambulatory and balancing somatic exercise

   - This stage of the process is very long. Depending on the mover, experience, and day of the experimental phase, this can take anywhere from 30 minutes to 1 hour.

Break: This break is intended to give the mover some time to recover from the affective and sensory overload that can result from the somatic exercises

4. Introduction of the concept
5. Static somatic re-priming/re-setting: sensory isolation, rhythm tracking, locating seat of intuition

   - This should be a resetting process and should only take between 5–15 minutes

6. Concept-specific static somatic exercises: seat of intuition, center-periphery, layers
7. Concept-specific dancing memory exercise
8. Concept-specific ambulatory and balancing somatic exercise that follows from the dancing memory exercise
9. Context-specific ground improvisation exercise
10. Ground improvisation pre-video review worksheet

   - This process should take around 30 minutes
   - The ground improvisation exercise should last between 3–5 minutes
   - Video recording should occur for stages 8 and 9

9b. (Optional) Aerial improvisation exercise
10b. (Optional Aerial improvisation pre-video worksheet

(*continued*)

(continued)

Large break: This break is intended to recover from the somatic and improvisational exercises. But more importantly, this break is intended to allow the mover to remove themselves affective from the embodied recollection of the improvisational movement. Because of the time constraint, we used this time for a long lunch.

11. Ground (and aerial) improvisation video review and post-video worksheet with time stamps

    - It is necessary to go through an select time stamps of significant movements to analyze for embodied metaphors as the mover reviews the video.
    - At a later point, during the experimental assessment, we discovered that it is ideal to leave the post-video review worksheet for one day to become fully separated from the felt sense of the movement but also not too distanced from the experience.

12. Embodied metaphor brainstorming table

    - This is typically the longest stage in the process. This stage could last more than an hour to brainstorm two movements. As the movers repeated this exercise, they became quicker at completing movement analysis.
    - The thesaurus is an indispensable supplement to this part of the process since verbal language rarely captures the exact idea of the movement and it is useful to be able to go through various synonyms of ideas to find ones that best fit the movement meaning.

13. Static somatic exercises: sensory isolations, rhythm tracking, center-periphery, layers, locating seat of intuition
14. Generic dancing memory exercise: not concept specific
15. Ambulatory and balancing somatic exercise

(*continued*)

(continued)

- This stage of the process is very long. Depending on the mover, experience, and day of the experimental phase, this can take anywhere from 30 minutes to 1 hour.

Break: This break is intended to give the mover some time to recover from the affective and sensory overload that can result from the somatic exercises

16. Introduction of the concept
17. Static somatic re-priming/re-setting: sensory isolation, rhythm tracking, locating seat of intuition

    - This should be a resetting process and should only take between 5–15 minutes

18. Concept-specific static somatic exercises: seat of intuition, center-periphery, layers
19. Concept-specific dancing memory exercise
20. Concept-specific ambulatory and balancing somatic exercise that follows from the dancing memory exercise
21. Context-specific ground improvisation exercise
22. Ground improvisation pre-video review worksheet

    - This process should take around 30 minutes
    - The ground improvisation exercise should last between 3–5 minutes
    - Video recording should occur for stages 8 and 9

9b. (Optional) Aerial improvisation exercise
10b. (Optional Aerial improvisation pre-video worksheet

Large break: This break is intended to recover from the somatic and improvisational exercises. But more importantly, this break is

(*continued*)

(continued)

intended to allow the mover to remove themselves affective from the embodied recollection of the improvisational movement. Because of the time constraint, we used this time for a long lunch.

23. Ground (and aerial) improvisation video review and post-video worksheet with time stamps

    - It is necessary to go through an select time stamps of significant movements to analyze for embodied metaphors as the mover reviews the video.
    - At a later point, during the experimental assessment, we discovered that it is ideal to leave the post-video review worksheet for one day to become fully separated from the felt sense of the movement but also not too distanced from the experience.

24. Embodied metaphor brainstorming table

    - This is typically the longest stage in the process. This stage could last more than an hour to brainstorm two movements. As the movers repeated this exercise, they became quicker at completing movement analysis. Some movements could be analyzed in as little as 5 minutes.
    - The facilitator proves essential during the beginning stages of this process. The facilitator is not there to share with the mover how they perceived the movement or which movements they thought were the significant ones. Instead, the facilitator facilitates the brainstorming process; they work as a sounding board for the mover to help them think through their ideas with clarifying and follow up questions to the mover's initial ideas.
    - The thesaurus is an indispensable supplement to this part of the process since verbal language rarely captures the exact idea of the movement and it is useful to be able to go through various synonyms of ideas to find ones that best fit the movement meaning.

Polished Inquiry Method

1. Static somatic exercises: sensory isolations, rhythm tracking, center-periphery, layers, locating seat of intuition
2. Introduction of the concept
3. Static somatic re-priming/re-setting: sensory isolation, rhythm tracking, locating seat of intuition
4. Concept-specific static somatic exercises: seat of intuition, center-periphery, layers
5. Concept-specific dancing memory exercise
6. Concept-specific ambulatory and balancing somatic exercise that follows from the dancing memory exercise
7. Context-specific ground improvisation exercise
8. Ground improvisation pre-video review worksheet

9b. (Optional) Aerial improvisation exercise
10b. (Optional Aerial improvisation pre-video worksheet

- To go straight through these steps of the process takes roughly 1 hour

Large break

9. Ground (and aerial) improvisation video review and post-video worksheet with time stamps
10. Embodied metaphor brainstorming table
11. Movement analysis
12. Question construction. Examples of this will appear in the results chapter.
13. Embodied metaphor/concept selection for further sequencing. This step will be discussed in more detail in the method commentary section.

Variations

While there is probably an infinite number of ways to go about applying this method, we have determined that there are four key approaches (Figs. 7.2, 7.3, and 7.4):

1. Concept. This would be to engage in inquiry on the basis of the abstract concept alone. So, the process would be only concerning a concept, e.g., freedom, and then further sequencing the threads of the embodied metaphors that are chosen. This process seemed to be the one preferred by each mover.
2. Context. In this situation, the mover would choose a context, e.g., police brutality, and engage the process this way. While the movers enjoyed having more specificity, they realized that this was sometimes too broad since they held many different kinds of ideas on the context. For this reason, one can do two phases—one for concept and one for context. And there are differences in results depending on the order you choose.
3. Concept then context. The idea is to do a short version of the process for the concept without completing full movement analysis and then introduce a context. This will allow the mover to focus on the concept from the perspective of the context. It is interesting to

Fig. 7.2 Constance doing improvisation exercises

Fig. 7.3 Shannon doing improvisation exercises

watch the way movement changes and stays the same when qualifying the concept with the context.

4. Context then concept. This allows the mover to get a general feel for their relationship to the context and then narrow in on a specific aspect. One can see whether or not the idea of the concept already emerged in the context-specific movement. The movers seemed to prefer context-concept over concept-context because it felt like a more natural progression of thought.

Choreographing Narrative Method

Once the mover discovers their questions regarding the theme, they can begin to think about their answers. If one were to come up with the question "why do I fear isolation?", they can begin the process of self-reflection. Whatever their response to their question is, that response will contain a number of concepts. For example, it might be that one fears isolation because isolation makes them face their true self, or isolation makes them feel abandoned, or isolation feels like not existing. Each one of the these responses has its own core concept. And usually there is either more than one response or further reasons to explain why the response is what it is. These responses make up the content of the narrative choreography.

For one to understand how to build an embodied metaphor, it helps to look at examples. I've included a large number below from Köveces,

Fig. 7.4 Shannon doing aerial improvisation

Johnson, and myself. Because movement is dynamic and has multiple layers, building embodied metaphors to communicate meaning is a matter of finding the appropriate intersecting primary metaphors to blend together over a short movement sequence. A dancing metaphor is not going to be a freeze frame of a move. It will include the movement from beginning to end. This means that there are many ways to temper one embodied metaphor with another. This is possible because a single movement phrase involves numerous forces, directionalities, heights, and intensities. Building a dancing metaphor follows the same process as analyzing a movement during the embodied inquiry phase, except in reverse. One might use a table and start with the concept and account for all of the various dynamics that can go into a movement to present meaning.

One important thing about building dancing metaphors is that neither movement nor meaning is static. This is helpful because meanings and ideas flow into, and build upon, one another in the same way that movements do. For this reason, dancing transitions become advantageous moments for communicating meanings along a narrative arc. One can use the movement transitions to lead from one idea to another to establish a train of thought, as it were. Transitions are what unites the movement sequences in a way that tells a coherent story. This means that embodied metaphor dynamics will be utilized for transition movements as well. Speeds and heights can communicate how the story is progressing, both in time and affectively.

While there is no set number of sequences in a movement piece, it is not uncommon for pieces to range from 3–6 minutes, if it is a stand-alone piece. These basic choreography formats typically allow for three to four significant movement sequences. Each movement sequence will be a constructed dancing metaphor, which means that there will be roughly five to six transitions, if accounting for intro and conclusion. Therefore, even a small piece will require a large degree of movement conceptualization.

There are three main points to keep in mind when constructing choreography in this way. First, this process is really only suited to narrative choreography. If one's goal is to entertain or showcase their newest skills, the need for embodied metaphors is absent. Second, even though dancing metaphors are intended to communicate objectively, no performance has to rely on them alone. Given the riches of our sensory perception, all of the other accentuations, such as lighting, costuming, and apparatus (if doing aerial) contribute to the sharing of meaning since each of these aspects also connect to our embodied metaphors indirectly: blue lights can help one convey that they feel "blue", which is associated with down and slowness. When layered on top of each other, slow, downward movement under blue lighting can effectively communicate sadness. The more intersections one can build, the higher the probability of communication—though to be sure, embodied metaphors can work on their own but even silence communicates meaning. This leads to the third point, which is that music is chosen after the fact so that it can punctuate the movement. Typically, performers start with a music selection and then choreograph in a way that accentuates the musical components and lyrics. Choregraphing dancing metaphors must work the other direction. Again, it is possible to start with a song, if this is the meaning one wants to communicate, and then build the dancing metaphors around it. But if the goal is to choreograph something personal that is not indirectly represented, music should be selected as a complement to the movement.

Example Embodied Metaphors
Kövecses
Metaphor and Emotion

Anger	*Fear*
Hot fluid in container: boiling	Fluid in container: filled with fear
Fire: slow burn, smoldering	Hidden enemy: crept up
Insanity: insane	Tormentor: tormented
Opponent: struggling with	Supernatural being: haunted
Captive animal: unleashed	Illness: sick with
Burden: carries around	Insanity: insane with
Aggressive animal behavior: snarl	Divided self: beside myself
Trespassing: draw the line	Opponent: took hold of me
Physical annoyance: pain in the neck	Burden: weigh heavily
Natural force: stormy	Natural force: engulfed
Functioning machine: get him going	Social superior: dictated by
Superior social: governed by anger	

Happiness	*Sadness*
Up: up	Down: brought me down
Being off the ground: six feet off the ground	Dark: dark mood
Being in heaven: heaven on earth	Lack of heat: cold, fire burned out
Light: brighten up	Lack of vitality: disheartening
Vitality: alive with	Fluid in a container: filled with sorrow
Warm: warmed	Physical force: terrible blow
Health: feel great	Natural force: waves wash over
Animal that lives well: pig in shit	Illness: heart-sick, heals
Pleasurable physical sensation: tickled pink	Insanity: insane with
Fluid in a container: overflowing with joy	Burden: staggered under
Captive animal: broke loose	Living organism: drowned with sorrow
Opponent: knocked out, overcome	Captive animal: got out of hand
Rapture/high: drunk	Opponent: seized
	Social superior: ruled by

Love	*Lust*
Nutrient: starved for	Hunger: starved
Journey: long, bumpy road	Vicious animal: beast
Unity of parts: as one, breaking up, inseparable, fused together	Heat: hots
Closeness: close	Pressure inside container: exploded
Bond: tie	Insanity: insane
Fluid in a container: overflowing with	Functioning machine: turn me on
Fire: burning	Game: first base
Economic exchange: putting more into this than you	War: conquest
Natural force: swept off	Physical force: knocked me off feet
Physical force: magnetically drawn	Natural force: waves
Opponent: fight	Social superior: ruled by
Captive animal: let go	
War: conquered	*Pride*
Sport: made a play	Fluid in a container: filled with
Illness: heart-sick	Superior: not let her
Magic: enchanted	Economic value: underestimate
Insanity: crazy about	Cause injury: injured
Social superior: ruled by	Cause physical damage: put a dent in
Rapture/high: high on	
Appetizing food: sweetie-pie	
Small child: baby	
Deity: on a pedestal, worship	
Valuable object: treasure	

Fire/heat	fiery, kindled, heat
Hunger/eating	satisfy hunger, fell starving, fall to food, appetites
Animal behavior	tigress, ferocity, hold back
War	battle, conquest
Insanity	raving maniac, madness
Natural force	flood, drowning
Rapture	drugging, dizzy with
Opponent	struggled, demolished control
Tormentor	torturing, tormented
Container	depleted, exhausted, exploded
Unity/bond	merged, united
Possessed object	let have, possessed, took
Physical force	jolts, magnetism
Game/play	playing games, play with
Magician	magic, spell, weaved around
Trickster	lure, bewitch
Social superior	overpower, driven by

Emotions

Internal pressure	Opponent	Wild animal	Social superior
Natural force	Trickster	Insanity	Fire
Hunger	Physical agitation	Burden	Mechanical force
Electric force	Gravitational force	Magnetic force	

Relationships

Sharing objects	Distance	Warmth	Bond
Economic exchange	Building	Implement	Machine
Plant	Journey	Valuable commodity	

Metaphor Appendix

Body is a container for the emotions	Emotional states are bounded regions
Emotional self is a divided self	Existence of an emotion is in a bounded space
Existence of emotion is possessing an object	Intensity of emotion is quantity/amount
Intensity of emotion is heat	Intensity of emotion is strength of effect
Negative emotions are illnesses	

Abstract effort is physical effort	Abstract existence is physical existence
Abstract function is physical function	Abstract maintenance is physical maintenance
Action is motion	Action is self-propelled motion

Affection is warmth

Anger is a burden	Anger is a burning substance
Anger is a hot fluid	Anger is a natural force
Anger is a social superior	Anger is a substance in a container
Anger is an opponent	Anger is excess of Qi in body
Anger is fire	Anger is in the heart
Anger is insanity	Anger is physical annoyance
Anger is the movement of Qi	Anger is trespassing
Angry behavior is aggressive animal behavior	Angry person is a functioning machine
Angry person is a pressurized container	

Attempt at control is struggle with force
Attempt at emotional control is trying to keep a complete object together
Attempt at emotional control is trying to hold back a captive animal
Attempt at emotional control is trying to overcome an opponent
Attempt at emotional control is trying to suppress fluid in a container
Control over action is control over self-propelled motion
Control over an emotional act is control over motion

Attributes are possessed objects

Bad things are cold	Bad things are dark
Bad things are down	Bad things are nonvaluable
Being bad is being low	Doing evil is falling
Evil is a force	

Behavioral responses are other-propelled motions
Cause of anger is a physical annoyance
Cause of anger is trespassing
Cause of changed state in emotion is motion caused by a force
Caused events are other propelled
Causes are forces
Causing harm to a proud person is causing injury
Causing harm to a proud person is causing physical damage to object
Changes are movements
Complex systems are complex physical objects
Complex systems are living organisms
Dynamic relation between two entities is a physical force acting on another
Human relationships are complex objects
Conceited person is up/high
Creation is making
Development is growth
Desirable is valuable
Desirable things are valuable commodities
Desire is hunger
Desires that control action are external forces that control motion
Difficulties are burdens
Difficulties are impediments to motion

Effect of an intense emotion is insanity

Skip section on emotions because it is just the general version of all the other emotions

Expected progress is a travel schedule

Experiences are objects

External events are large, moving objects

Fear is a burden	Fear is fluid in a container
Fear is a hidden enemy	Fear is a hidden object
Fear is a natural force	Fear is a social superior
Fear is a supernatural being	Fear is a tormentor
Fear is an illness	Fear is an opponent
Fear is insanity	

Friends are containers (that open to each other)	Friendship is a journey
Friendship is a living organism	Friendship is a machine
Friendship is a possessed object	Friendship is a strong bond
Friendship is a structured object	Friendship is a valuable commodity
Friendship is an economic exchange	Friendship is closeness
Friendship is fire	Friendship is warmth

Communications between friends is sharing objects (innermost experience)

Communication is sending

Good things are light	Good things are valuable
Good things are warm	Being good is being upright

Happiness is a captive animal	Happiness is fluid in a container
Happiness is a natural force	Happiness is pleasurable physical sensation
Happiness is an opponent	Happiness is being in heaven
Happiness is being off the ground	Happiness is healthy
Happiness is insanity	Happiness is light
Happiness is rapture	Happiness is up
Happiness is vitality	Happy is light
Happy is up	Happy is warm
Happy person is animal that lives well	Being happy is being off the ground

Ideas are valuable commodities

Important is central

Increase is growth	Increase is intensity

Intensity is amount/quantity Intensity is strength of effect	Intensity is heat

Interaction is an exchange	Relationships are buildings

Internal is external
Intimacy is closeness
Lastingness is strength

Life is a journey	Life is a play

Purposeful activities are journeys	Purposes are destinations

Loss of control is loss of control over a strong force

Love is a bond	Love is a captive animal
Love is a disease	Love is a fluid in a container
Love is a journey	Love is a natural force
Love is a nutrient	Love is a physical force
Love is a social superior	Love is a unity

Love is a mutual economic exchange	Love is an opponent
Love is closeness	Love is fire
Love is insanity	Love is magic
Love is rapture	Love is a sport/game
Love is war	Love is a physical force

Lust is a game/play	Lust is a natural force
Lust is a physical force	Lust is a social superior
Lust is a substance in a container	Lust is a trickster
Lust is a unity	Lust is a vicious animal
Lust is an opponent	Lust is wildness
Lust is fire	Lust is heat
Lust is hunger	Lust is insanity
Lust is magic	Lust is pain
Lust is pressure inside a container	Lust is rapture
Lust is war	Lustful person is a functioning machine

Marriage is a physical or biological unity
Meanings are objects
Means are paths
Mental activity is manipulation
Mind is a container

Morality is resisting	Morality is strength
Nonphysical harm is physical damage	Nonphysical unity is physical unity
Object of love is a deity Object of love is a small child Object of love is appetizing food	Object of love is a possessed object Object of love is a valuable object Object of lust is possessed object

Passive experience are the physical effects of forces
People are containers
Person out of control is a divided self

Pride is a fluid in a container	Pride is a superior

Pride is an economic value
Progress is movement

Sad is dark Sadness is a burden Sadness is fluid in a container Sadness is a living organism Sadness is a social superior Sadness is an opponent Sadness is a burden	Sad is down Sadness is a captive animal Sadness is a lack of vitality Sadness is a physical force Sadness is an illness Sadness is insanity
Shame is a burden Shame is a fluid in a container Shame is a blocking of the world Shame is having no clothes Shameful is a person divided	Shame is a decrease in size Shame is an illness Shame is hiding away from the world Shame is physical damage Shameful person is a worthless object
Surprise is a natural force Surprised person is a burst container	Surprise is a physical force

Vanity is an indulgent person

Example Embodied Metaphors
Johnson
Embodied Mind, Meaning, and Reason

Understanding is grasping
Purposes are destinations
Causation is forced motion
States are locations
Change of state is motion
Causes are forces
Hindrances are obstacles to motion
Actions are self-propelled motions
** Thinking is motion

Mind is a body
Thinking is moving
Ideas are locations
Reason is a force
Rational thought is motion
Unable to think is unable to move
Line of thought is a path
Thinking about X is moving in the area around X

Thinking is perceiving
Thinking is object manipulation
Thinking is preparing and eating food

Examples
Shay

Following Johnson's example:

Thinking is moving	*Thinking is dancing*
The mind is a body	The mind is a movement
Thinking is moving	Thinking is dancing
Ideas are locations	Ideas are pointed gestures
Reason is a force	Reason is gliding
Rational thought is motion	Rational thought is transitioning
Being unable to think is being unable to move	Being unable to think is being grounded
A line of thought is a path	A line of thought is a leap
Thinking about X is moving in the area around X	Thinking about X is a Fouette
Communicating is guiding	Communicating is reaching toward the crowd
Understanding is following	Understanding is a Grand Jeté
Rethinking is going over the path	Rethinking is repeating the sequence

Aerial Metaphors

Sadness:	performed close to the ground, closed body position
Anger:	flexed feet, kicking/punching transitions, upward direction
Devastation:	heavy wrapping draped slowly
Loss:	open wraps with downward direction
Ruination:	fast vertical drops from heavy wrapping
Tragedy:	slow drops, ooey gooeys, remaining in wraps
Trapped:	transitioning from heavy wraps to other heavy wraps, dramatic wrapping
Confused:	transitioning from heavy to free to heavy wraps, slow and unsmooth transitions
Fighting but losing:	top to bottom transitions or rotations downward, moving between various kinds of heavy wraps
Freeing oneself:	either finding ways of dropping heavy wraps with one swoop or quick but liquid unwrapping
Triumph:	dramatic, pointed, liquid unwrapping while moving upward in direction

Freedom:	wide body shapes, working on outside of apparatus
Travel:	moving from outside to outside, large climbs, switches
Twisted shapes:	negative emotions (trapped, confused, manipulation)
Spirals and cartwheels:	positive emotions (joy, love, romance, travel)
Slow, multistage drops:	negative or uncomfortable life or emotional transitions
Slow roll ups:	overcoming
Fast roll ups:	jubilance, success
Vertical slack drops:	(often) negative emotions (finally broken, no return, surrender, finality); obviously these can be accentuated or modified to represent positive emotions, but they need to be punctuated with positive music or gesture polish.

Worksheets

- Examples of completed worksheets will be included in the results chapter.

Pre-video Reflection

Concept/Topic:
Movement sequence:
Chosen apparatus:
Prior to video review:
General reflections:

____

____

Significant affective and emotional responses:

____

____

Significant movement recollections:

____

____

Post-video Reflection

Post-video review
General reflections:

____

____

Identifiable affective and emotional responses:

____

____

Identifiable significant movement dynamics, rhythm, and shapes:

____

____

Identified Primary Embodied Metaphors

____

____

Selected Primary Metaphors to sequence next

____

____

Order of Primary Metaphors in next sequence

____

____

Identifiable critical questions about topic from sequence

____

____

____

____

Embodied Metaphor Brainstorming Table

Timestamp	*Shape*	*Tempo*	*Force or intensity*	*Levels or height*	*Mood or tone*	*Emotion*	*Kinesthesia*
–							
–							
–							

See Fig. 7.5.

Fig. 7.5 Shay, Constance, and Shannon doing worksheets

Some Commentary

As I've mentioned, the order, time, and cueing of the process will depend on various factors, such as professional training, length of time in the process period, movement experience, somatic familiarity, etc. In my experiments, I worked with each mover for three full days. This was a very long and arduous process for each mover, and especially me. The movers found themselves to be very emotionally exhausted and I found myself to be exhausted from the amount of attention and focus I had to invest to monitor both the facilitation and the recording of the mover's responses and feedback. As I've also mentioned, the process with the first two movers went painfully slowly and we did not get to the level of completion that I was able to achieve with the other two movers. But this was because the first two movers were actively developing the process with me and were helping test out various approaches to doing the exercises, whereas the other two movers were more providing experiential input and feedback regarding their ability to develop these skills on their own through the training process. People who attempt to engage in this method will have a similar experience even thought the method is provided. This is because it will still require some trial and error and basic practice to enact the method in a reaonable amount of time.

Because the somatic work is paramount to the method, all breaks and interruptions will require a somatic "resetting", which allows the mover to get back into the embodied-affective place necessary to engage in the movement process. This resetting will depend on the amount of time or degree of interruption, such that a short break will only require a very small resetting—a couple of minutes—whereas a long break or a large interruption will require repeating the somatic exercises from scratch. When the resetting is short, the mover can simply reset herself by going to the core concept-specific static somatic exercises. If the method stops for a snack break, then one can engage in a quick version of the process: 1 minute kinesthesia, 2 minutes concept-specific sensory modalities, 3 minutes static dancing memory, proceed into movement-based exercises. This is extremely helpful when the mover needs more breaks between exercises to recover or to engage in dialogue with the facilitatory about how the process is coming along.

Stopping should not, in itself, be regarded as an interruption. One important insight that emerged is that stopping during the method, whether when static or when in movement, is very informative and can be helpful in moving forward. As one dancer from Nagrin's article reported:

> Sometimes you will find yourself stopping for any number of reasons: confusion, uncertainty, fear or just plain physical exhaustion. There's nothing wrong with that, *as long as you deal directly with the stopping*. When stopping becomes part of the dance, it has its own excitement and meaning. (Nagrin 1994, 159)

Stopping often turns out to be a result of feeling stuck, as it were. Sometimes the mover can lose their affective attunement, sometimes they can lose themselves in space, sometimes they can get drawn to one specific sensory stimulation to the point of falling out of the somatic exercise. And most significantly, the mover can be in the middle of movement and feel like they don't know "what to say" or "lose their thought" in the middle of a "sentence". All of these moments are places of embodied experience that is telling. Instead of the mover feeling as if they have failed at the exercise or feeling as if they need to start over, the mover must, instead, allow these moments and, if useful, take account of them. Embodied sensory and affective responses very rarely indicate a lack and resting or stopping is a response. Therefore, the mover should be encouraged to not resist or suppress the urge to pause or follow a sensation. These responses are natural and should be expected and the mover should allow themselves to go into those places and explore—through their kinesthetic capacities—why the pause occurred and whether the pause was necessary.

The process can be as extensive as the mover wishes to make it. The ways in which the mover can select, analyze, and further explore movement sequences is almost endless. One can select only one sequence, use the main sequences of the movement piece, or take one or more sequences and continue the process in relation to their corresponding questions, concepts, or related contexts. Each movement sequence can be analyzed to such an extent that one can produce something like a logic tree. All of these decisions depend upon how long the mover's narrative piece will be, how focused the mover wants their narrative to be, or, practically, how much time they have to choreograph.

There are a few moments in the process that can call the authenticity of the movement into question, even though the process is purposefully from a subjective perspective and so cannot be "checked". First, there is a question of authenticity or, rather, of success in the method if the movement within the video aligns almost directly with the mover's personal style of movement. If the video displays repeated signature movement patterns or flares, then the mover can possibly be said to either not be in the somatic kinesthetic-affective state of mindbody or they are merely on movement autopilot. In these cases, the process will need to be restarted from the beginning at a later time. A second point where questions of authentic subjectivity can come into question is during the embodied metaphors worksheet exercises. One might suppose that the mover is or could or might or unknowingly be adapting the movement analysis so as to find the metaphors that they think belong to the movement. What this amounts to is assigning the movement meaning rather than uncovering its meaning. This is a hard question because it is impossible to know when or if this is happening, unless it is just an egregious and apparent rejection or rushing of the movement analysis process. However, beyond outright "cheating", the tweaking of metaphors is not a problem to such an extent that it undermines the process. For one, it is likely impossible, or at least highly improbable, for one to not purposefully intervene in the movement analysis with propositional content. The movement analysis worksheets are not Ouija boards; they do not guide the hand sans any considered input. If the mover has an inkling of what the movement means but cannot seem to be able to get the "right" movement analysis, then it is sensible for them to intervene. What this tells us is that the process is a fully subjective experience and subjectivity carries knowledge in many ways. If the mover knows her movement to have a meaning and attempts to "fudge" the worksheets, they are not failing in the analysis, they are simply more self-reflectively aware of the movement meaning.

Because the embodied critical inquiry method is fully subjective and narrative choreography, in some cases, aims for objectivity, it holds that the movements from one will not match up to the movements of the other. The movement sequences that manifest during the movement exercises are subjective and so will have readability and significance primarily to the mover herself. Therefore, the narrative choreography, should not utilize these movements extensively. The purpose of the critical inquiry is to engage in self-reflection to draw out one's questions about the world. The choreography is intended to communicate those answers to others.

Our subjective embodied metaphors are as complex as we are and are deeply intimate. Therefore, the mover will need to build their narrative embodied metaphors from glaringly cliché movements that are constitutive of our embodied shared understandings.

Bibliography

Brodie, Julie, and Elin Lobel. *Dance and Somatics: Mind-Body Principles of Teaching and Performance.* Jefferson: McFarland and Co., Inc. Publishers, 2012.

Chaiklin, Sharon, and Hilda Wengrower. "Introduction". *The Art and Science of Dance/Movement Therapy: Life Is Dance.* Chaiklin, Sharon and Hilda Wengrower (eds.). New York: Routledge Press, 2016: xix–xxxii.

Fraleigh, Sondra. "Dance Maps: A Guide for Dance Experiences". In *Moving Consciously: Somatic Transformations through Dance, Yoga, and Touch.* Fraleigh, Sondra (ed.). Urbana: University of Illinois Press, 2015: 213–226.

Nagrin, Daniel. *Dance and the Specific Image: Improvisation.* Pittsburgh: University of Pittsburgh Press, 1994.

CHAPTER 8

The Movers: Results

In this chapter, I include the bits and pieces of my experiments that help clarify some of the aspects of the method. The most significant feature of this chapter are the completed worksheets. The worksheets just are, in essence, the results. The worksheets all look different. This is intentional. They were, and still are, fully experimental. That they are different is how I use them to capture the "results" through them. There are some major changes in the worksheets at the beginning that seem to peter out towards the end. This is because I had almost no idea how to best organize the process or the recording of information when I started and was finally able to smooth out most of the wrinkles by the end. What, in the end, takes roughly an hour or two to accomplish, took about ten hours to accomplish at the beginning. Each person attacked the worksheets differently and with different styles; some used bullet points, some were narrative, some were handwritten, and sometimes I had to help out. I have attempted to consolidate and organize them somewhat to make their presentation smoother. I have not included the partner worksheets because they follow the same pattern as the others. The truly unique aspect of the partner work was the effectiveness of the somatic work. I include a set of reflective remarks from each mover at the end of their section. Their prompts were to just talk about their experience doing the method and to mention anything they thought stood out. I have left these remarks in their own words.

One will be able to see major and subtle changes over the worksheets. There is one particular part of the worksheet that gets cut out most of the

S. Welch, *Choreography as Embodied Critical Inquiry*,
https://doi.org/10.1007/978-3-030-93495-8_8

time since we never had enough time to go as far as we wanted to. If one were to go from zero to full choreography using this process, it would take a few weeks. If one were to just use these worksheets, which would then call on each part of the worksheet, it might take something like a week to go as far as one might need to go. There's also another part that moves from verbal remarks to just movement time stamps—the movement sequences said all that there was to say at that point. In other places, one can see the worksheets get longer since we learned to keep track of even the weirdest ideas. Some worksheets do not include every part. Sometimes we did not get far enough to go all the way through, sometimes someone wanted to focus on only on aspect. As I've mentioned, this process is not rigid; it is intended to facilitate and organize one's process rather than to dictate the process. I do not include every worksheet that each person did. That seems a bit of an overkill. I chose the ones that I thought were interesting, either in terms of the organization or in terms of the content. I make sure to note which day, concept, apparatus, embodied metaphors, and final questions the worksheets are about above it so that one can know what they're looking at to begin with and do not need to hunt for the information.

Had it not been for poor Connie and Shannon's love and devotion to the art, and their faith in me as a decently functioning person, I would not have been able to pull this process together into something uniform enough to call a method. This was genuinely some of the hardest intellectual work I've ever done in my life, and they surely carried much of the burden.

CONSTANCE

Fig. 8.1 Constance

Constance Echo Palmer moved to the United States from Germany where she continued her classical dance training, focusing on ballet. Later she discovered circus arts, primarily performing in small stage productions and at corporate events as a magician's assistant, fire performer, and aerialist. Constance's formal training includes aerial arts, movement theatre, Pilates, and somatics and has a B.S. in Biology with an Anatomy and Physiology emphasis. She has owned the nationally hailed studio, The Space, where she hosted the (copywritten) Open Grip movement artists' draft showcase and the nationally acclaimed Atlanta Aerial Arts Festival, which hosted world renowned movement theorists and performing artists. She recently sold her studio to run away with the circus, literally, as the star and rigger for the Lewis and Clark traveling circus. Highly motivated and versatile, Constance can perform on stage or work behind the scenes in tech and rigging. In 2015 she began touring with a Rock and Roll Cabaret in which she was the lead rigger and circus artist regularly performing 4 aerial or fire acts per day. Constance is a rigger and rope access technician, so safety and professionalism are of the highest importance to her craft (Fig. 8.1).

Day: 1
Concept: Sexual Pleasure
Definitive embodied metaphors: relaxation, enjoyment qua release
Questions: We did not get this far

Concept/Context: Sexual Pleasure
Movement Sequence: Ground Improv 1
Apparatus: Ground

Prior to video review:

General reflection: (What was it like as the mover?)

- blooming
- growth
- sunshine
- expanding
- expansiveness
- images of different flowers
- bright

Significant affective and emotional responses:

- expansiveness
- relaxed
- open
- flowy
- instinctive

Significant movement recollections:

- expanding from contracted shapes
- reaching upward and outward

Post-video Review:

General reflections:

- reaching
- stretching
- expanding

(*continued*)

(continued)

Identifiable affective and emotional responses:

- calm
- blissful
- relaxed

Identifiable significant movement dynamics, rhythm, and shapes:

- movement patterns repeating: small hops and rolling movements
- flowey
- asymmetrical body shapes
- many changes in level

Concept/Context: Sexual Pleasure
Movement Sequence: Aerial Improv 1
Apparatus: Sling (Low Flying)

Prior to Video Review:

General reflection:

- There are so many more external distractions now than for the ground improv.
- The simple presence of the sling is also kind of a distraction.
- Pressure to get into the sling but purposely waiting.
- Noting the importance of environment for my creation process.

Significant affective and emotional responses:

- more free to movement/flight is freeing
- sling is hugging/embracing
- morphing of concepts from blooming and growth to carefree

(*continued*)

(continued)

Significant movement recollections:

- more free to float, swing, twist, and hang
- kind of drapey
- I enjoyed the moments of release from off center without a decided path or direction

Post-video Review:

General reflections:

- can visibly tell when the concept started to feel different: changing
- shift in movement from growing/expanding to swinging and floating
- looked much more internal, personal, and even unintentional at times

Identifiable affective and emotional responses:

- the rolling, small shapes, and tucking made it look and feel more personal, internal
- this is not a performance for others

Identifiable significant movement dynamics, rhythm, and shapes:

- repetition of rolling shapes and swinging takes me back to the first idea of the day: cycles and the swing dance
- draped over the sling in different orientation stands out to me

Brainstorming Session:
Shape/Movement:

- low in the sling
- moving between open/wide and tucked/small shapes
- rolling in the sling
- swinging

(*continued*)

(continued)

Tempo:

- slow
- not always even
- moments of release to swing

Intensity/Force

- low force
- variable intensity

Embodied Metaphors:

- comfort
- being at peace
- **state of relaxation**
- life cycles
- day cycles
- **release and enjoy/freedom**

Day: 2
Concept: Violence
Embodied metaphors: hopelessness, fueled, isolated, chased
Questions: We did not get this far

Concept/Context: Violence
Movement Sequence: Ground Improv 1
Apparatus: Ground

Prior to Video Review:

General reflection:

- high energy
- led by extremities
- chaotic and quick
- could have stayed in somatic moment watching "the dance"
- shoulders and feet felt energized and leading the way

Significant affective and emotional responses:

- exhausted
- defeated
- perseverance
- failure

Significant movement recollections:

- bursts of movement/energy
- fight alternating with falling
- grounded moments are defeated

Post-video Review:

General reflections:

- looks like a battle or a possession
- fighting something invisible
- interesting to see myself attack the floor

(*continued*)

(continued)

Identifiable affective and emotional responses:

- pain
- empathy for the mover (watching from the outside)
- looks mostly like a losing battle with burst of perseverance

Identifiable significant movement dynamics, rhythm, and shapes:

- the collapses onto the mat were more impactful watching it whereas the pull to movement by my feet and shoulders felt stronger/more important while in the movement

<u>Brainstorming</u>

Shape	*Tempo*	*Force/ Intensity*	*Levels*	*Mood/Tone*	*Emotion*	*Kinesthesia*
Large open star	Explosive/ fade	Decreasing	Pop ups/ bursts	Combative/ trial	Perseverance	Stretchy
Folded, compressed, crushed	Dead stop	Fizzled	Lowest	Isolation	Defeated	Tight

Identified Primary Embodies Metaphors:

- flames, fires, embers, feeding the fire w/breath
- pressure being applied, being forced closed, being acted upon
- **hopelessness**/isolation

Selected Primary Metaphors to sequence:

- fueled
- hopelessness

Concept/Context: Violence
Movement Sequence: Aerial Improv 1
Apparatus: Fabric

Prior to Video Review:

General reflection:

- violent
- tumultuous
- struggle
- trying to get up the silk
- spent a lot of time fighting the silk and where it is in relation to self

Significant affective and emotional responses:

- anger
- frustration
- exhaustion

Significant movement recollections:

- general struggle
- struggle to get up/climb
- struggle to get the silk on or off or out of the way

Post-video Review:

General reflections:

- physical exhaustion from the movements
- felt my breath a lot
- repeated folding/crushed shapes from floor improv
- nothing was ever accomplished/completed

Identifiable affective and emotional responses:

- overwhelmingly, struggle and frustration
- followed by exhaustion

Identifiable significant movement dynamics, rhythm, and shapes:

- kicking
- thrashing

(*continued*)

(continued)

- throwing
- fighting
- messy

Brainstorming

Shape	*Tempo*	*Force/ Intensity*	*Levels*	*Mood/ Tone*	*Emotion*	*Kinesthesia*
Loop recoil	Snap, Punctuated	Low force/ low intensity	Bottom of the top	Steadfast	Driven	Flexed/ engaged
Veiled, squatty	Paused	Heavy resistance	Middle	Somber	Confused	Effortful

Identified Primary Embodies Metaphors

Loop Recoil:

- being chased/not caught
- breaking of a bond
- abandoned or exit
- holding in place and releasing
- committed
- happening around me/not by me

Open Silk Hand Push:

- bodybag/fleshy
- trapped in sight
- suppressed
- restricted
- concealed
- compressed
- vacuum packed like meat at the grocery

Selected Primary Metaphors to sequence:

- being **chased** and not caught
- **insulated**

As one can see, the worksheets started out as a disaster. But Connie is a lover of science and really liked the idea of the worksheets. So, we worked diligently on how best to use them. She was the person who developed the brainstorming table. This was the most grueling part of the process.

She had three initial concerns. First, she was worried that the process would be subjective, in every way and in every part. We discussed how all persons are socially situated and so even people who engage in a "scientific" method would not be able to do so "impartially". Another point we talked about was how each of us always has our own ways of thinking inside our head (body) and that we know what we mean all the time and that we just have to find a way to suck all the life out of our ideas to communicate them effectively to others. Claims are always subjective; people just tend to word them in a way that obscures that. But more importantly, we talked about how this process was an embodied way of having a dialogue with yourself and that another's ability to comprehend what was going on the inside is irrelevant. But this is when I had to think through what the difference between subjectivity and objectivity in movement and reading embodied metaphors really entailed. Her second concern was whether the external factors would be detrimental to conducting the method. This we learned through experience. Over time, we realized that we needed to learn to address external factors of all kinds as part of the symphony of lived experience rather than as intrusions to perception. This became a key introductory discussion point with each of the movers and seemed to be effectual. This turned out to be very helpful when Deena and I got hit with a surprise torrential thunderstorm. Her third scientist question was whether there was even a way to measure success. How would we know, she asked, whether or not she had done it right. There are two answers to this question. The theoretical answer is that you know when you hit it. That embodied metaphors can be read intuitively seems like a fluffy claim until you're in the midst of it. Every single time a mover would find the right term for the movement, they would absolutely and unquestionably know it and they would stop immediately. There was never a time where a mover didn't feel sure they had captured what their movement was saying. A practical response is that the method is successful as far as it is productive. If the method helps you in your choreography process, it's successful; if it doesn't, it's not. Luckily, all movers felt like the method was successful on both accounts.

Constance's Remarks

My participation in this project was really interesting and exciting because it combined a lot of different fields that I work in, as well as different approaches to movement. I am an aerialist with a background in dance that works in touring shows and coaches master classes. I also have a background in different fields of somatic practice such as yoga, Pilates, and laban movement analysis.

The process of working through the method step by step and subsequently creating improvisations was energetically taxing at times. The physical components of the somatic and improvisational exercises were not themselves tiring, but the mental involvement was exhausting. In other words, depending on the environment, it could be more or less difficult to focus on the exercises at hand and/or feel ready to move. Elements such as temperature, size of the space, and implementation of physical boundaries affected my movement, although sound was the component that affected me the most. For example, if abrasive music played, or chatter was present in the space during the somatic exercises, it was much more difficult to drop into the flow-state. In addition, it would take me much longer to reset at the start of the following improvisations.

The success of the improvisational activities was dependent on the success of the somatic exercises. Basically speaking, the more successful the somatic exercise was (in terms of clearing my mind and dropping into the body-mind) the more freely I was able to improvise. Although the context and concept, depending on which we were working with at the time, greatly influenced the quality of my movement during the movement improvisations, they were definitely more meaningful if the somatic exercise was successful. I personally felt that dealing with a context made the concept much more approachable and created more meaningful movement during the improvisations. For example, the tone or mood of the concept or context definitely affected the types of movements as well as my personal emotional response to the movements as I performed them, as well as when I reviewed them.

Working with a partner in these research and movement activities was really intriguing to me. I have previously worked with dance partners on choreography and improvisation, but this was the first time I worked through the method in this book with a dance partner. It was fascinating to observe when we contrasted and complemented one another in shape and spatial arrangement both on the ground as well as during aerial movement improvisations. The person that was selected for me to work with was a colleague and someone that I know quite well. I do wonder if working with a stranger would yield similar results and movement segments that were equally interesting and meaningful.

Overall, I found the method we used to create movement segments to be a very detailed and involved process, which was quite successful. The movement segments were successful when they appeared meaningful in the post creation review process, in which we analyzed videos of the improvisational work. I am grateful for the opportunity and the space to perform this research. Additionally, I'm thankful for the guidance through this method. I look forward to delving deeper into the method to discover what I can create through it. My greatest curiosity is whether or not this method can be employed in solitude or if a facilitator is necessary because I found the guidance through the somatic exercises and the selection of the concepts and contexts by a third party to be quite invaluable at the time.

Shannon

Fig. 8.2 Shannon

Shannon Stanley is a classically trained ballet dancer. Starting at the age of 3, she began training and performing with companies such as the Central Pennsylvania Youth Ballet, Virginia Ballet Theater, Governor's School for the Arts, and the Miami City Ballet. Several years ago, she took a hiatus from dance while getting her Bachelor's in Humanities from Florida State University. Wanting to use her classical training and learning a new art form, she discovered and fell in love with aerial arts after seeing photos of local artists. Ms. Stanley continues to expand her knowledge of fitness and the performing arts and is excited to share her passion with others. These days, Shannon is a professional performing aerialist, an elite aerial coach, and a circus choreographer. Shannon choreographs aerial for Marvel Universe Live and the national circus Circus Smirkus. She has also taught choreography workshops choreography at NECCA (New England Center for Circus Arts) (Fig. 8.2).

Day: 2
Concept: Beauty
Embodied metaphors: struggling to maintain, escape from suppression, never ending switchbacks, curious revelation

Questions:
- Why is beauty unstable?
- How can anyone be free to be beautiful under so much pressure?
- How can we escape external pressure of beauty?

Movement Sequence: Beauty Concept
Chosen Apparatus: Stretchy Split Fabric

Prior to video review:

General reflections:
I wanted to be wrapped because I wanted that warm tight feeling. I felt the need to be more vertical than spinning or horizontal movements or swings. I had to explore different grips and wraps until one felt right for what I was searching for.

Significant affective and emotional responses:
Security, excitement, timid, warm, anticipation

Significant movement recollections:
Multiple wraps, bouncing and rebound, ascending, hanging with minimal to no wraps

Post-video review:

General reflections:
2:38
4:25
Frequent changing from spaztic to multiple wraps. Bouncing and repetitive "open" and "closed" positions. Wrapping over the head and neck (unusual) frequent use of flamenco grip and uneven grips. Unusual way of locking wrists.

(*continued*)

(continued)

Identifiable affective and emotional responses:
Uncomfortable, safe, embarrassment, hiding, looking for warmth, reaching up, playful

Identifiable significant movement dynamics and shapes:
Multiple wraps, swiftly going in and out of movements that are opposite shapes, wrapping up slowly but unwrapping fast, hanging unevenly

Identified Primary Embodied Metaphors
1. Burrow
Struggling to conformity
Struggle for consistency
Disconnected
Struggling to unite
Struggling to maintain

2. Finding least resistance
Funneling
Break free
Caper
Sudden escape from suppression

Selected Primary Metaphors to sequence next
Struggling to maintain
Sudden escape from suppression

Aerial brainstorming

Shape tempo force levels mood emotion kinesthesia
2:38: squiggly fast high force middle constrained suppressed. tense then not squeezed

Water finding least resistance
Funneling
Break free
*caper
Sudden escape from suppression

4:25 cuddle pose static high force middle effortful grasping tense
Half Tucked
half scrunched

(*continued*)

(continued)

Burrow
Struggling to conformity
Struggle for consistency
struggling to unite
struggling to maintain
"Why is beauty unstable?"
"How can I be free to be beautiful under so much pressure?"
"How can we escape external pressure of beauty?"

Movement sequence: Beauty Ground Improv
Chosen apparatus: Dance

Prior to video review:

General reflections:
Movement was a little awkward and hard to find physical bearings. Body was being pulled off balance and kept "jumping" through idea. Had to allow weight to drop in order to release into idea.

Significant affective and emotional responses:
Aggravation, anticipation, excitement, happiness, fear

Significant movement recollections:
Jumping, rolling horizontal, weight dropping and bouncing on mat

Post-video review:

General reflections:
:50
2:48
Repetition. A lot more shoulder influenced than noticed. A lot of up emphasis and horizontal movements. Sweeping legs in a circle.

Identifiable affective and emotional responses:
Being pulled. Frustration. Stumbling, bursts of energy.

Identifiable significant movement dynamics and shapes:
Skips back, arms in circles, pushing opposite arms, wrapping arms around head. Shoulder circles.

(*continued*)

(continued)

Identified Primary Embodied Metaphors

1. First time roller coaster
Mountain switchbacks
Exploratory journey
Blind switchback corners
Boomerang
Never ending switchbacks

2. Pixie
Surprisingly being attracted to something
Surprisingly drawn to something
Moth to flame (with different ending)
Ethereal
Going towards something surprisingly
Discovery
Curious revelation

Selected Primary Metaphors to sequence next
Never ending switchbacks
Curious revelation

Brainstorming: beauty concept ground

Shape. tempo force levelsmood emotion kinesthesia
1
:50 squiggle quick accent (v) grounded Confused lost fluid

First time roller coaster ride
Mountain switchbacks
Exploratory journey
Blind switchback corners
Boomerang
Never ending switchbacks
"Why is my attitude towards beauty sometimes good and sometimes bad?" "Why am I ambiguous about beauty?"

2: plinko light low to mid pleasant surprised fluid
2:48 flutter

(*continued*)

(continued)

Pixie
Surprisingly being attracted to something
Surprisingly drawn to something
Moth to flame (with different ending)
Ethereal
Going towards something surprisingly
Discovery
Curious revelation
"How is it still possible for me to be (pleasantly) surprised by beauty?"

Day: 2
Context: Beauty and YOU
Embodied metaphors: labored conformity, compliance, snubbing, jealous cynicism

Questions:
- What do I get from compliance with beauty expectations?
- Do I actually snub beauty standards?
- Why are beautiful people considered more valuable than ME?

Movement sequence: ground improv beauty context
Chosen apparatus: ground

Prior to video review:

General reflections:
My body wanted to be horizontal driven by my feet. Every bone in my feet needed to be articulated. Within being horizontal, there needed to be some sort of shield or somewhere to be under or simply face down.

Significant affective and emotional responses:
Failure and attempt to retry, balance, frustration, tension, relief,
Significant movement recollections:
Rolling through each part of feet, making circles and cyclical motions, "threading" through other body parts, horizontal balance positions.

Post-video review

General reflections:
1:00
3:05
Fingers mimicked foot articulation. Long lunges trying to "flip" body position putting head under legs. One legged horizontal balances were often unsuccessful but immediate reattempted.

(*continued*)

(continued)

Identifiable affective and emotional responses:
Obsession, scared, hiding, intrigue, unknown

Identifiable significant movement dynamics and shapes:
Folded and threaded places. Foot "moonwalking " rolling through feet repeatedly, one legged balance with leg in back attitude bent position.

Identified Primary Embodied Metaphors
Mirroring
Laborious conformity
Actual mirroring
Concerted effort to be cooperative
Intentional cooperation
Cooperation with external expectations
Compliance with meeting expectations
Turning towards external expectations
Going through the motions of meeting expectations

Selected Primary Metaphors to sequence next
Laborious conformity
Compliance with meeting expectations

Identifiable critical questions about topic from sequence
What do I get from compliance With beauty expectations?

Brainstorming beauty context ground

Shape. Tempo. Force. Levels. Mood. Emotion. kinesthesia
pathway Intensity (height) tone
1
1:00 intersecting static. high force low. Purposeful. Obligated. Torquey
triangles high pressure Twisting
Focus on negative space

Laborious Conformity
Metaphorical Mirroring

(*continued*)

(continued)

2
3:05 flipping consistent low key low compliant cooperative torqued high force Regular

Actual mirroring
Concerted effort to be cooperative
Intentional cooperation
Cooperation with external expectations
Compliance with meeting expectations
Turning towards external expectations
Going through the motions of meeting expectations
"What do I get from compliance with beauty expectations?"

Movement sequence: Aerial Beauty Context Improv
Chosen apparatus: Sling

Prior to video review:

General reflections:
A strong need to be leaning and weight held in the middle part of body (chest/ belly/ pelvis) tension in the lower body, feet in particular. Wanting to contort upper limbs and press into an open fabric and wrap into it. At the very end felt the need to get out as soon as possible.

Significant affective and emotional responses:
Frustration, uncomfortable, tension, hurt, confused,

Significant movement recollections:
Pressing into open fabric and rolling. Pressing against the fabric. Leaning, rolling up, threading between poles, changes in tempo

(*continued*)

(continued)

Post-video review

General reflections:
2:17
3:55
Pushing body into open fabric. Leaning middle body into fabric. Getting fabric wrapped around one specific body part. Emphasis on looking straight up and "gazing" down.

Identifiable affective and emotional responses:
Sadness, fatigue, frustration, determination, fight or flight

Identifiable significant movement dynamics and shapes:
Wraps around specific body parts, arching shapes, arms behind back or behind head

Brainstorming aerial context

Shape. Tempo. Force. Levels. Mood. Emotion. kinesthesia
pathway Intensity (height) tone

1: UPSIDE DOWN static high force middle sneaky dodgy taut
Vertical high intensity shady locked out
fabric moving
Inverted pencil
Asymmetrical

2:14

Dodge
Side-eyes
Judgmental
Homer fades into bush

Nothing to see here
Peace out
Judging the expectations
Turning away from the expectations
Rejecting the status quo

(*continued*)

(continued)

Side-eye the status quo
Being skeptical of the status quo
Scoff the status quo
Sneer at the status quo
THUMBS DOWN BITCHES
Snub the status quo
"Do I genuinely snub beauty?"

2: arched high force low uptight annoyed tense
Outside looking feeling bad pushing
but no fucks
3:55

Pushing away from others
Escaping the need for acceptance
Hateration
Drinking the hater-aid
Jealousy driven cynicism
"Why are beautiful people even considered more valuable than ME?"

Shannon's initial questions were entirely different from Connie's. For Shannon, she was concerned that is would be impossible to engage in an authentic form of movement when professional movers have extremely trained and habituated bodies. She was not sure that, even if she could find her somatic place—which she was skeptical about, since she left ballet precisely for this reason—that her embodied affective experience could forge with her kinesthetic dance experience. A lot of our initial discussions focused on how authenticity in movement didn't mean that you were going to get the "right" result or the "authentic" movement. The idea is more concerned with allowing yourself to be open to how your body wants to move you rather than squashing that movement because it might not have good lines. It also means that you have to be open to being moved at all without having planned or thought about the movement prior to the exercises. This may have had more to do with the fact that she is both a ballerina and an aerialist. I say this because her final question was about how she would be able to do aerial in an authentic movement manner without opening herself up to serious injury; aerial has a strict vocabulary for a reason and that reason is safety. One must necessarily go through a series of movements in order to achieve a safe position or movement. Here, we talked a lot about how training becomes important. Just as embodied cognition theorists address one's ability to make deliberate decisions during expert real-time action, an aerialist with strong training can know where she's going and what she's doing and know how best to do those things even when doing so while being moved. As it turned out, and much to her surprise, her authentic aerial movement required almost no complex movement that would lead up to any particular technical "skill". For her, it ended up being about doing and undoing wraps in certain ways with certain dynamics rather than doing wraps to "get somewhere". Even more surprisingly, she found that she didn't feel compelled to get too far off the ground and, at times, she felt no need to get off the ground at all. It turns out that the mere embodied relationship to the apparatus could help one articulate an embodied idea without actually needing to get "on" it.

There were five very significant points of feedback that she provided during the process that helped me better understand what to look for and how to explain things theoretically. The first was how very vulnerable—but therapeutic—the whole method made her feel. At many times she felt very overwhelmed and felt that, by the end of it, she had learned a lot of new things about herself which she had purposefully blocked off. This made perfect sense since much of the somatic method stems directly from dance movement therapy; and I modeled my role as facilitator on the principles of dance movement therapy. The second insight is that she found her somatic-based improv movement to have no correlation to her personal style or her professional training. There was a lot about the videos that shocked her to the point that she didn't want to see it because it looked "gross". It was at this point that I realized that that was one way to test whether or not one was truly moving largely from an embodied knowledge place: the more the video looks like your trained, personal style, the less likely it is that you're somatically engaged. The third is that the time stamps proved to be paramount to being able to manage the brainstorming worksheet. Prior to using the time stamps, the worksheet was pretty much pandemonium. Fourth, she really felt like she needed a buddy. Initially she thought that it was because she was more of a mover and less of a worder; but really, in the end, it just turned out that it was really helpful to have someone throw out suggestions and alternatives when one gets stuck. This led to another central component, which is the thesaurus.... because even the buddy gets stuck. Once we told ourselves it was ok to go down the thesaurus rabbit hole, the whole process got swifter. Moreover, by giving oneself permission to use the thesaurus (when others might poo poo the idea) is that it gives the mover a way to more easily do the process on their own. Once Sorsha (mover #4) got the hang of how I was using the thesaurus to throw out suggestions, she took over completely.

Shannon's Remarks

I am a classically trained ballet dancer that took a turn to aerial performance and coaching. I don't incorporate yoga and meditation into my everyday practice and chose to go with kickboxing and weightlifting to supplement my aerial training. Working in a somatic headspace to create improv was both intimidating and a little funny sounding to me.

I was confident my ground improv would be authentic, but the real question was whether the aerial improv could be authentic. With aerial work, there are many other safety measures that need to be addressed and prioritized. The apparatus and the rigging I'm hanging from is a large concern but can be inspected before performing. Technique is a top concern to help protect my body from injury. An incorrect body position that leads to incorrect muscle engagement can lead to strains, pulls, and tears. But when working on fabric, or silks, I am always thinking of my wraps and the order in which they are laid. I want to be 100% sure that my wraps are correctly sequenced so they can hold my weight. An incorrect sequence of wraps can lead to falls with catastrophic bodily harm. How was I able to completely remove all of these fundamental ideals that I had been trained to focus on? And for good reason! I also worried about my natural muscle memory. Movements and sequences I have practiced thousands of times that are engrained into my body that I could perform in an instance. I was not sure if I would be able to release everything I knew to safely improv in the air.

I quickly found that outside visual light stimulation distracted me from finding my somatic place. I was the most comfortable laying in a position on my stomach where I could create a dark visual space with my arms wrapped around my face. Once I was able to find what I believed to be my "soma" it became easier every session to find it again. In the end, I could reset and be ready to improv in about five minutes.

My solo session's prompts were chosen differently. For my first session I chose my prompt from a group of crumbled papers. The one I chose at random was "isolation". For my second prompt, Shay decided to give me a prompt that she knew would be personally controversial, beauty. This process proved to be both emotionally draining and therapeutic. So much that one day we had to take a break and eat an entire pineapple upside down cake with our hands! I had to dig into the details of personal experiences that inspired me to move. The moments that were joyous felt good to celebrate but I also had to explore the negative memories that the prompts dug up.

It was imperative that Shay was there to help guide the completion of the work sheets. I was capable of telling my stories through my movements but completely unable to verbally express what I was conveying. I wasn't fully aware of the questions that were being expressed until she would offer insight on what they may be. At some points the accuracy would bring tears to my eyes. I had no idea I was digging that deep into my inner being while in my improv. As I was viewing video of my improv directly after, the actual content of my movement was coming from a far deeper place than I had originally realized.

When it came time to do our partner improv, it is important to note that the only reason I would even agree to do a partner improv was because it was with a person I trusted. My partner had been a longtime friend and current training partner, so I was able to be completely open to the task. It wasn't until we were completing the worksheets and discussing our experience that my partner and I had actually pulled memories from the same day! It is in my opinion that had my partner, and the facilitator for that matter, not been trustworthy friends, I do not believe I could have been as authentic as I was. I believe I could have completed the exercises successfully, but it would have been reserved and not authentic to my true self.

Deena

Fig. 8.3 Deena

Deena Skolnick Weisberg is Assistant Professor in the Department of Psychological and Brain Sciences at Villanova University. She earned her B.S. in Symbolic Systems from Stanford University and her Ph.D. in Psychology from Yale University. At Villanova, Weisberg directs the Scientific Thinking and Representation (STAR) Laboratory, which researches scientific thinking in children and adults, the development of imaginative cognition, and the roles that the imagination plays in learning. Weisberg is also the co-director of the Galápagos Education and Research Alliance (GERA), a partnership between universities and Galápagos community members to promote science and conservation in the Galápagos Islands (Fig. 8.3).

Weisberg is also a current and founding member of Tangle Movement Arts, an all-female aerial dance theater troupe based in Philadelphia, PA. Tangle's work is devised collaboratively by its ensemble, reflecting individuals of diverse identities with a particular emphasis on queer and female experience. Weisberg has performed regularly with Tangle since the company's inception in 2011, and her work on trapeze and aerial silks has been featured in the company's popular Fringe Festival shows.

Day: 3
Concept: Renewal
Embodied metaphors: whirlwind, inquisitiveness, questioning

Questions:
- What inquisitiveness compels us to constantly seek after renewal?
- Will fulfilling a quest lead to renewal?

Concept/Topic: Renewal
Movement sequence: Improv 1
Chosen apparatus: Ground

Prior to video review:

General reflections:
I wanted to fly, to take off, to leap, soar, swoop. It felt like I had to launch, get up in the air. I wanted to stay on my toes, throw my hands up and out, try to get as big as possible in the space. At one point I stopped to see if I could access the part of the memory where I was sitting on the ground but it didn't feel as good/right—I felt like I had to reach out to the world and welcome it in. I wanted to feel the rush of the air as it moved past my body.

(Shay said to stop using imagery but this will help me remember: It was snowing on me as I was lying on the ground, big puffy flakes. I could feel them tapping on my arms and legs as they landed. I felt giddy, like a kid with a day off of school.)

Significant affective and emotional responses:
Dizzily joyful. Giddy. Open/accepting.

Significant movement recollections:
Moving forward, moving up, not wanting to stay still, a restlessness in my body to find the next thing and the next thing.

(*continued*)

(continued)

<u>*Post-video review*</u>:

General reflections:
Totally had forgotten about the rocking motion I did on the ground—that was the part where I thought I wanted to be on the ground but actually I wanted to be higher/more dynamic
As before, it felt like I was moving faster than I was, but this concept definitely got out more big-body movements, circles, waves, higher levels

Identifiable affective and emotional responses:
Happiness, wonder, curiosity, welcoming-in

Identifiable significant movement dynamics, rhythm, and shapes:
00:21–00:25
1:53–2:03
3:11–3:23
Arm circles, cycles/waves of motion

Concept/Topic: Renewal
Movement sequence: Improv 1
Chosen apparatus: Silks

<u>*Prior to video review*</u>:

Why pick that apparatus:
I had to get higher! Climb to the ceiling! Also the waves that the silks could make (and later discovered the sound of their rustling as they moved on the ground)

(*continued*)

(continued)

General reflections:
I loved being able to climb and to have the silks move with me. I wished it was possible to just suspend myself in the air and work with them (but I did what I could with physical limitations)…to the point where I really wanted to climb but it felt wrong to have the silks be under tension and just static poles.

Significant affective and emotional responses:
Joy, curiosity, exploration, childishness, wonder

Significant movement recollections:
Restlessness, exploration/seeking after novelty, feeling joy in the sensations (touch/sound) as the silks moved

How is embodied communication on apparatus different, better, worse than ground:
In this case, the aerial improv felt much more satisfying than being on the ground—I loved being able to get higher, to incorporate the movement of the silks, to feel them move in big waves around me and gather them in. I wanted to be surrounded by them and keep them moving as much as possible

Post-video review:

General reflections:
The things that I was noticing and responding to as I was doing the improv were mostly the motion and texture of the silks and how they sounded on the floor—those moments were a little hard to see in the video because they relied somewhat on my point of view and how close I was, but there's still some good waves in here, which is what I'd wanted.

Identifiable affective and emotional responses:
Curiosity, exploration, collaboration, playfulness, silliness

Identifiable significant movement dynamics, rhythm, and shapes:
00:16–00:21
1:46–1:50
3:10–3:16

(continued)

Brainstorming

Timestamp	*Shape*	*Tempo*	*Force/intensity*	*Levels (height)*	*Mood/tone*	*Emotion*	*Kinesthesia (state or position of the muscle/body)*
00:16–00:21	Circle, wave, whirlpool, swirly	Fast, punctuated	Forceful, force is being applied, propelled	Mid, chest-height	intensely purposeful, searching, relentless	Motivated, driven, enthusiastic, curious/seeking	Bouncy, waving, loose but not slack, fluid

(continued)

Curious fluid swirls relentlessly propelled

Whirlpool
Shay: "We need a curious whirlpool"—add in exploring/seeking
Going down a rabbit hole
Exploration
That swirl at the beginning of the yellow brick road
Wanderlust
Adventuring (voyage?)
Flurry
Zephyr
Zephyrous
Adventure-seeking
Whirlwind/vortex
Inquisitive

Inquisitive whirlwind

Questing

Does renewing require a quest?
Will the quest lead to a renewal?
Is renewal automatic or must it be sought?
Fate means it's automatic but destiny requires a fulfilling
Is it one's destiny to renew?
Does destiny always result in renewing?
Does destiny require us to constantly reinvent ourselves?
How will we know when the quest is fulfilled?
Is renewal the result of a successful quest?
Are we compelled to constantly seek after renewal?
What compels us to constantly seek after renewal?
Is renewal a form of reinvention—purposeful reinvention?
Will fulfilling the quest lead to renewal?
How is renewal a quest?

Identified Primary Embodied Metaphors
Quest

Whirlwind
Inquisitive

(*continued*)

(continued)

Identifiable critical questions about topic from sequence
What inquisitiveness compels us to constantly seek after renewal?
Will fulfilling the quest lead to renewal?

Selected Primary Metaphors to sequence next
Inquisitive

Attempt to build embodied metaphor for inquisitive:

What compels us to constantly seek after renewal?
Will fulfilling the quest lead to renewal?

What movements do this?
Climbs—explosive upward movement
Pointed/flexed toes
Butterfly climb (but don't fold all the way over)
Big Russian/Russian flirt climb—but careful about not being flirty.
Keep feet very flexed

Start with components that are not negotiable for the concept—that must be there for the expression
Upright
Facing forward
Movement (not static)—but what part is moving?

(*continued*)

Concept	*Shape*	*Tempo*	*Force/ intensity*	*Levels (height)*	*Mood/tone (less specific than emotion, what others might pick up on)*	*Emotion (more specific, how you would describe yourself as feeling)*	*Kinesthesia (state or position of the muscle/ body)*
inquisitive	Upright, forward-facing, upward spiral	Punctuated, steadily, medium fast	Not at lot of force, but intense	Climbing / upward-directed, expressing upward momentum	Positive (not perky), satisfied, anticipatory	Joyful, excited, eager	Flexion (point to flex), tension like a coiled spring (potential momentum/energy?—prepared to spring, compressed/ jack in the box)

Because I had worked out a lot of the kinks at this point and was able to explain the goals and steps of the method, theoretically and practically, Deena didn't have a lot of questions to begin with. She was open to just going through the process and then simply reflect on it at the end. In addition to the reflections she provided, she ended up giving me a list of further questions to respond to at the end. Deena did, however, provide a lot of pointed feedback. And since I had gotten through most of my "how did this work" questions beforehand, she was able to offer more feedback on her lived experience doing the process. But we did still have plenty of new things about the method to discover. First, the worksheets I included are on a concept that she chose rather than one that I introduced on the spot. For this we learned that it was harder to use one's own concept than to do the process with a surprise concept. This is because one has plenty of time to construct a propositional understanding of the idea and then try to force that knowledge into the embodied knowledge during the movement exercises. So it required more somatic work to avoid doing so. The second is that using an explicitly negative concept can be almost debilitating to the point that the method becomes unworkable. We used the concept "rejection" on the first day and it was almost too heavy. But I suppose this is just the way life is and one will have to be emotionally prepared to take these kinds of somatic dives. Another discovery was just how effective the dancing memory exercise could be, if done properly. Previously, I had introduced it in various ways at various times in the various tested orders of the exercises. While it always made sense, it didn't always seem to be especially fantastic. We were able to improve on this by linking the dancing memory exercise to the improv exercise without any break in between them. Doing so allowed me to develop the instructions to "find the motion" in the memory to the "find the motion of the memory" to "dance with the motion of the concept" to "dance the concept". Starting with the dancing memory exercise and then slowly building to dancing the concept using improv made the conceptual improv exercise feel more natural.

Here is the one question she asked me that I thought had been lingering all along and my response to her:

Question: The ground and aerial improvisation sessions themselves generated (what I thought were) interesting insights about the concept. For example, as noted above, my "Commitment" improvisations brought about the realization that my commitment is about trust and freedom and connections to other people (even though I'd thought it would be about feeling weighed down and overburdened). Given this, why not use these insights directly to tell us about the concept? What would be different about the method if this were a part of the process, instead of then translating aspects of the movement from the improv into questions?

Response: I have also thought about this. But really, the embodied metaphors that come out of the improv exercises don't really tell us much. So if the concept is "Renewal" and the embodied metaphor is "inquisitive", you don't know what, exactly, the relationship is between inquisitiveness and renewal to build a choreography from it. Like, you can't just say "renewal is inquisitiveness". It doesn't say how or why. So building the question allows you to figure out how you actually want to explore the relationship. The connection between them could be infinite. You, personally, will only have a few ways that you see the relationship. So building the question gives you the ability to explore that relationship so that you can have the content needed for the choreography piece.

Deena's Remarks

I started off rather skeptical that I would be able to do Shay's method properly. I wasn't skeptical of the idea itself—I'm on board with the idea of embodied cognition, otherwise I wouldn't have volunteered to help out in the first place—but rather of the thought that I personally would be able to experience my embodied knowledge through movement. I'm a highly verbal, analytical, intellectual person, and I was worried that I just wouldn't be able to access a somatic way of knowing.

I was wrong. It worked. Turns out that being a body in the world for 40 years really does lead to some knowledge.

We did some initial somatic warm-ups, which felt fairly natural to me from having done aerials and yoga. After discussing that experience, we did a somatic reset and Shay gave me my first concept: rejection. I had an immediate and spontaneous reaction throughout my body—my whole face clenched up, I couldn't take a full breath, my muscles started tensing. It felt utterly awful. I had tears running down my face all through my initial standing experience of the concept, and I'm pretty sure I cried through most of my floor dance improv too. My neck muscles were sore the next day from the tension I'd been holding. (I'd focused on the aspect of the concept that involved me being rejected; at no point whatsoever did I consider the aspect of the concept that would involve me rejecting something/someone else, so there's an interesting tidbit for a therapist to chew on…)

Shay asked me to bring up a memory of rejection and then to move with the movement of that memory. I think she'd been worried that that would be an odd request but (at least for me, at least for this concept), that made sense right away. I was remembering a conversation in which I'd offered help at a personal cost to me but that offer was rejected, and that conversation had a push-pull aspect to it that translated straightforwardly into movement.

Afterwards, when I watched the video and selected a moment to explore more deeply, I eventually was able to translate it into a question about how the fear of rejection might make me lose part of my identity, and that was the first moment that I felt the revelation: I formed the question and immediately thought, yes! That's where the fear and the destabilization and anxiety come from, the thought that rejection might damage my view of myself, my ability to be the person I think I am. I didn't know it (explicitly, verbally), but my body knew it. It worked. Holy shit.

The second day we worked on commitment, and then commitment in the context of mentoring. Working on the concept of commitment was incredibly illuminating to me because it went in a totally unexpected direction. I'd initially expected commitment to feel terrible too, because of all of the commitments (and over-commitments) I have in my professional and personal lives: I should be writing, I have to be a teacher and mentor and researcher and mom and all of these other things that grind me down… I was expecting my exploration for commitment to feel heavy and burdensome and mundane and awful. But it didn't. When I started to move with commitment, it felt amazing. It felt freeing. Rather than feeling like a burden, my commitments felt like hands holding me up in a trust fall—my commitments were the connections to all of the people and ideas in my life that gave me purpose and made me feel safe. I'd never before thought of my commitments like that, but once I did, it made perfect sense: yes, my commitments are about feeling connected and feeling trusted and feeling safe.

Again: It worked. Holy shit.

A general note I'd like to add at this point is that my physical explorations of the concepts/contexts brought up a lot of mental imagery (e.g., rejection was like orange globs in a lava lamp, commitment made me feel like a silver tree with branches stretching out in all directions). Shay encouraged me away from describing these aspects of my experience, I presume so that I would stay more focused on the somatic components of the concept rather than the imagistic components, but those aspects felt to me like they also held important insights about the concepts/context.

A second general note is that I think I had a somewhat easier time working with a concept and then moving with a memory of that concept than I did with doing the method with a concept in a context. That could be because I only did that once during our time together and just wasn't used to it, or because I was having trouble feeling a deep connection with that particular combination of concept/context, etc. But my suspicion is that finding a memory that linked to the concept already provided a context, and then adding a different context on top of that was just too much to work with and inflected the concept in a different way than what I was expecting. Whatever it was, I didn't feel that the one concept+context improv I did was particularly successful or provided particular insight, but that's just one experience so take it with a grain of salt.

On the third day, Shay let me pick my own concept; I chose renewal. It was a little strange to have the concept not come as a surprise; I was anticipating it and so perhaps didn't enter into movement as easily as I did the other two days because my brain was doing too much pre-gaming. I don't think that damaged the eventual exploration, just something interesting to note.

Working with this concept also didn't go the way that I thought it would. I thought it would bring up memories of springtime or birth/rebirth but that wasn't the case at all. The memory I settled on had lots of moments of stillness, yet my movements couldn't be still—I needed to rock back and forth, to climb up, to wave the silks around like sails, etc. I felt strongly in all of the improvs I did, but these ones perhaps most clearly, that there was a way that I was supposed to be moving, that felt right for the concept, and I couldn't fight against it or do something different because that felt wrong. For example, for renewal, I really wanted the silks to be swirling around and billowing, but climbing them put too much tension on them so I had to let go.

This was also the concept that we worked on the most to bring it through the entire process, which was a really enjoyable part of the experience—it was fascinating to see how I could translate the linguistic concept into movement and then to translate the movement back into a linguistic question and then to translate that question into bits of choreography. I'm going to keep working on this, moving into my next performance, and see where it takes me.

Sorsha

Fig. 8.4 Sorsha

Sorsha Masters was born and raised in Smyrna, GA. She was exposed to the arts and movement practically since birth. Her wonderful parents exposed her to dance-centric musicals at the age of 3, and from that point on, she knew she wanted to be a performer in some capacity, but her preference was to be a dancer. At the age of 4 she was enrolled in dance lessons at a local studio where she started her education in the genres of ballet, tap, jazz, musical theatre, lyrical, and contemporary. She has performed for Atlanta Ballet and Gotta Dance Atlanta under some of the top choreographers and instructors in Atlanta. Early in 2018 Sorsha began training in aerial arts at The Space in Atlanta. On top of being a trained dancer of twenty years, Sorsha has also received acting training from the Alliance Theatre in Atlanta. After receiving extensive acting training, she went on to perform in several plays in the Marietta and Atlanta areas. Sorsha is now currently a dance instructor in Dallas, GA, but still resides in Smyrna where she choreographs and performs around the Smyrna, Marietta, and Metro Atlanta areas (Fig. 8.4).

Day: 3
Concept: Freedom
Embodied metaphors: relaxed, surrendered

Questions:

- Is there Gratification in surrender? Does freedom culminate in gratification? If I'm gratified does that mean I'm free?

Concept/Topic: Freedom
Movement Sequence:
Chosen apparatus: Ground

Prior to video review:

General reflections:
Definitely a concept I have never considered to try and communicate through movement.
It was interesting to feel what my body's natural response to the concept was.

Significant affective and emotional responses:
It felt almost joyous at times, but like it couldn't reach full joy because as we have all been taught; freedom comes at a cost.

Significant movement recollections:
Very open movements
Lots of back bends and a couple of reasonably high legs.

Post-video review:

General reflections:
04:42—Slow arm open archy thing
5:03 ish—Slow collapsing thing
5:37 ish—leggy, layout, reachy thing

Identifiable affective and emotional responses:
Open and almost joyous at times, but it also felt like there was tension in it.

Identifiable significant movement dynamics, rhythm, and shapes:
Lots of open arms, lots of bending backwards, and lots of wide shapes.

(*continued*)

Brainstorming							
Timestamp	*Shape*	*Tempo*	*Force/intensity*	*Levels (height)*	*Mood/ tone*	*Emotion*	*Kinesthesia (state or position of the muscle/body)*
4:42	"Y" That's opening	Quick ish	High force/High Intensity	Low to high	Confident	Euphoric	Extension
5:03 ish	Fold to Upright Triangle	Steady	Low force/Mid intensity	Low to High	Hopeful	Gratified	Droopy

(continued)

Sweeping
Stretch
Elongate
Reveal
Stretch
Softening
Relax
Surrender
Release

Is freedom associated with surrendering? Does one have to surrender in order to become free? How can surrendering grant someone freedom? Can you become free through surrendering? What must you surrender to be free? Does gratification require a refusal to surrender? How can giving something up give me more freedom? Should other people surrender freedoms for my own gratification? Is there freedom in surrender?

Is there Gratification in surrender? Yes because there is freedom in surrender, and it is gratifying to surrender.

Wilt down/Bloom up
Flower
Following the sun
Accomplishments
Fulfillment
Crowning
Peak
Culminating

Does freedom culminate in gratification? If I'm gratified does that mean I'm free?

Concept/Topic: Freedom
Movement Sequence:
Chosen apparatus: Lyra

Prior to video review:

General reflections:
Definitely an unusual experience trying to communicate freedom on a rigid apparatus.

Significant affective and emotional responses:
Felt open and almost friendly.

Significant movement recollections:
Lots of arches, back bends, and hand movements.

Post-video review:

General reflections:
2:01 ish—Splitty layout ish thing
2:44 ish—Resting neck on lyra

Identifiable affective and emotional responses:
As a whole it seemed fairly warm. Almost happy.

Identifiable significant movement dynamics, rhythm, and shapes:
Lots of layouts and arm flourishes.

My time with Sorsha was the most surprising. I went into the experiment assuming I would do a poor job facilitating and that I couldn't live up to the standards of someone who was a full time classically trained dancer. She was curious about how this somatic work would happen because she hadn't heard of it before. Turns out, I was a perfectly fine facilitator and she was perfectly familiar with somatic work by another name. As such, our three days of going through the method was smooth sailing. She didn't have any questions because, I like to believe, I finally understood what I was doing enough to properly explain it to her. We had a fully lovely time. And this was the first time anyone was able to just go through the worksheets without me at all. I was really surprised by how smoothly and quickly it all worked out to be by the end. I learned with Deena that this process isn't only for the professional performer, and I learned with Sorsha that this process is not only for the academic. I was please to find, at the end, that this process seemed to be helpful or interesting or fun to all of the movers that occupied some position on the line between academic and performer.

As per the worksheets, we learned something new. I included here the worksheets for the concept Freedom since this was the most abstract concept I used; that is, this is the concept that I introduced that had the least obvious personal connection to the mover so that I could see how well this process would function philosophically. Ultimately, this kind of concept proved no more difficult than the others, which is what I hoped would happen. But what did happen for the first time is that a series of related concepts emerged that resembled something like a logical relationship. Instead of getting distinct embodied concepts, there were concepts that led into each other. It was unclear what was happening at first until we were trying to construct the questions. Sorsha wasn't finding any of the ones we put forward as suitable until I suggested that they were a continuous idea and not separate ideas. This immediately struck her as correct. This is how we ended up with questions that led into each other. The question we didn't get to attend to was how that would work in terms of sequencing them further.

Sorsha's Remarks

As a dancer I was trained to go against my intuition which is the exact opposite of what this method and practice is. Through this method I learned to engage my intuition and welcome what my intuition has to say instead of tuning it out. Through this method I learned how to engage my senses and welcome everything (smells, sounds, touch, etc…). Honestly, up until I started this practice, I had no idea how to experience movement without blocking out my surroundings. The first day it was a very thought intensive process to isolate senses, bring them together, and focus them into one place. Once I adjusted to the method, I realized that I had done similar things in the past, but it never quite made sense the way that this exact method made sense.

The second day for me was really different from the first. In my time on day one of my guided somatic setting/priming time it was very loud due to renovations happening where we were. On the second day, as we were doing more guided priming, I realized that a quiet room is not quiet at all. Since there was no renovation happening, I could really tune into ALL of the sounds that you normally don't think of as loud, and my god the world is loud! Honestly though, it helped me find rhythms in my movement when we got to the improvisational movement portion of things.

I can definitely say after having done the method both on the ground and in the air; it is very different and honestly more difficult to use embodied metaphors and really move as the idea of "Dreaming" or "Freedom" when you are in the air. As a dancer you are completely free to move as you please, because you aren't restricted by safety measures. There were things my body definitely WANTED to do in the air, but unlike being on the ground I had to stop and think about how if I were to do this certain thing, I could end up getting hurt due to safety restrictions when you're on an apparatus. I also realized that when I'm working with an idea on the ground, I tend to keep my eyes shut so that I can tune into my intuitive side more easily. That is not the safest thing to do when you're in the air on an apparatus, because sight is very much your friend when you're dealing with a sling or lyra, or both in my case.

The last day of practicing this method was almost a fully self-guided, priming day. I must say it was quite simple to "set myself" on the last day with minimal to no guidance. It became less of something I had to think intently about, and it became more of a true meditation of becoming in tune with my body. The last day wasn't a walk in the park though. After completing my movement analysis, I was trying to come up with the answer to my question that we came up with by using the information gathered on my brainstorming table. Unfortunately, I couldn't come up with the answer to my own question. After getting terribly frustrated, we came to the conclusion that I simply didn't have enough information to answer my question. Following that discovery we analyzed more of my movement, filled out more of the worksheet, and from that, I was able to expand upon my initial question and find my answer. I am the type that will try to solve a problem at all costs, but I learned through this process that sometimes you have to get more information to be able to solve the problem. I was genuinely surprised at how easy the "light bulb moment" came after I gathered more information. Throughout this whole process, everything that was said to me made complete sense, and putting it into practice gave me a sense of almost astonishment of the method. It's one thing to hear it and say "Ah yes, what you're saying makes sense.", but it is a whole other creature to go through this process, have some ups and downs, and come out on the other end with it making even more sense than it did when we started.

In my findings on my own since I've gone through this process It seems easier to be completely clear in the messages I try to convey within my work as a choreographer and as a dancer. I've been continuing work with the brainstorming exercises to find exactly what the message is that I'm trying to convey and how exactly I need to convey it.

Index[1]

[1] Note: Page numbers followed by 'n' refer to notes.

S. Welch, *Choreography as Embodied Critical Inquiry*,
https://doi.org/10.1007/978-3-030-93495-8

The manufacturer's authorised representative in the EU is Springer Nature Customer Service Centre GmbH, Europaplatz 3, 69115 Heidelberg, Germany. If you have any concerns regarding our products, please contact ProductSafety@springernature.com

Printed and bound by CPI Group (UK) Ltd, Croydon, CR0 4YY
30/09/2026
02225254-0003